The Ancestry of Domestic Cattle
Reprinted From The Twenty-Seventh Annual Report of the Bureau of Animal Industry

by US Dept. of Agriculture

with an introduction by Jackson Chambers

This work contains material that was originally published in 1912.

This publication is within the Public Domain.

This edition is reprinted for educational purposes
and in accordance with all applicable Federal Laws.

Introduction Copyright 2018 by Jackson Chambers

Self Reliance Books

Get more historic titles on animal and stock breeding, gardening and old fashioned skills by visiting us at:

http://selfreliancebooks.blogspot.com/

Introduction

I am pleased to present another title in the "Cattle" series.

The work is in the Public Domain and is re-printed here in accordance with Federal Laws.

As with all reprinted books of this age that are intended to perfectly reproduce the original edition, considerable pains and effort had to be undertaken to correct fading and sometimes outright damage to existing proofs of this title. At times, this task is quite monumental, requiring an almost total "rebuilding" of some pages from digital proofs of multiple copies. Despite this, imperfections still sometimes exist in the final proof and may detract from the visual appearance of the text.

I hope you enjoy reading this book as much as I enjoyed making it available to readers again.

Jackson Chambers

CONTENTS.

I.—SOME PROBABLE ANCESTORS OF DOMESTICATED CATTLE.

	Page.
Introductory	187
Historical sketch of the Bovidæ	188
The genus *Bos* and its five subgenera or groups	190
Taurine group—subgenus Bos	191
Bibovine group—subgenus Bibos	191
Leptobovine group—subgenus Leptobos	192
Bisontine group—subgenus bison	192
Bubaline group—subgenus bubalus	192
Extinct species in relation to ancestry of domesticated cattle	192
Pliocene epoch	193
Pleistocene epoch	194
Recent period	197
Origin of the modern breeds of cattle	212

II.—EARLY HISTORY OF CATTLE BREEDING, AND CLASSIFICATION OF MODERN BREEDS.

The cattle of the ancient civilizations of Asia	214
The cattle of Egypt	215
Carthaginian agriculture	216
Cattle of prehistoric Europe	216
The old Stone age	216
The new Stone age	216
The Bronze age	218
Origin of the cattle of modern European countries	218
Greece	218
Italy	219
Gaul	220
Iberia	220
Switzerland	221
Russia	221
Germany	222
The Netherlands	222
Scandinavia	224
Denmark	224
France	224
Great Britain	225
Ireland	226
Channel Islands	227
Origin of the principal types of cattle in America	227
Classification of modern breeds	228
A classification of British cattle	231
A classification of French breeds	231
Classification of breeds in America	232
Bibliography	233

ILLUSTRATIONS.

PLATES.

	Page.
PLATE XIII. The banting	192
XIV. Fig. 1.—Augsburg painting of the urus. Fig. 2.—Herberstain's picture of the urus	200
XV. The Vaphio cups and their scrolls	200

TEXT FIGURES.

FIG. 8. Diagram showing geological position of the Bovidæ	190
9. Skull of *Bos elatus*	194
10. Skulls of Algerian buffalo and *Bos acutifrons*	195
11. Skull of *Bos nomadicus*	196
12. Frontal aspect of four types of skulls	197
13. Skull of *Bos primigenius*	199
14. Skull of *Bos frontosus*	202
15. Skull of *Bos longifrons*	203
16. Skull of *Bos brachycephalus*	208

THE ANCESTRY OF DOMESTICATED CATTLE.[1]

By E. W. MORSE,

Specialist in Animal Husbandry, Office of Experiment Stations, U. S. Department of Agriculture.

I. SOME PROBABLE ANCESTORS OF DOMESTICATED CATTLE.

INTRODUCTORY.

The origin of our domesticated cattle has been the special object of study of many investigators, but the results of their labors have been written largely in foreign languages and have been published in widely scattered periodicals. In this article an attempt is made briefly to review and correlate the progress made by the many zoologists, paleontologists, anthropologists, and historians who have worked on this difficult problem, as the original publications are inaccessible except to those living in the vicinity of large libraries.

Many details which would be of interest to the zoologist have been omitted because it is presupposed that the majority of the readers of this article will be men interested in cattle breeding, who desire to obtain a general survey of what is known concerning the ancestry of European and American cattle but who are too preoccupied with other duties to read a mass of osteological details that would be both interesting and necessary to the scientist who is making a special study of the evolution of species. For those who care to pursue the subject further, a bibliography of some of the more important contributions to the subject is appended. The reader should also bear in mind that many facts regarding the ancestry of cattle are still unknown. Hence, those who are in haste to have all questions settled once and for all will derive but little cheer from the following pages. Much work remains to be done before final conclusions can be reached, but there may be some who, like the author, wish to know of the progress which has already been made, and for such the present work has been prepared.

By the term "cattle" we usually mean domesticated bovine animals, principally of two species—*Bos taurus*, European cattle, and *Bos indicus*, the humped cattle of India and Africa, commonly called

---

[1] The two papers comprising this article were delivered, in a somewhat abbreviated form, as lectures at the graduate school of agriculture at Ames, Iowa, in the summer of 1910. The author desires to acknowledge his indebtedness to the writings of Lydekker, Rütimeyer, Werner, Nehring, Wilckens, Keller, and Dürst, in particular, and to many other authors, some of whom are referred to in the text.

the zebu. The older writers in England used "cattle" or "cattell" as a collective name for all kinds of live animals held as property or reared to serve for food or beasts of burden, and the term sometimes included horses, sheep, swine, and by some writers even bees and poultry. Bovine animals were then designated as "horned cattle," and still more recently as "black cattle" and "neat cattle." "Black cattle" was probably first applied to the black breeds of Scotland and Wales. Later it had a more general application. "Neat cattle" were so designated because they were useful, "neat" being derived from the Anglo-Saxon word "neótan" (to make use of). The word "cattle" is another form of the word "chattel" and "capital," meaning originally goods or property, cattle among many primitive peoples being the most valuable goods, and frequently the measure of value of other kinds of property. The old English equivalent for cattle is "kine" or "kyan," derived from cy, the plural of cu, Anglo-Saxon for cow. The term "ox" is often used for cattle in general, but in a restricted sense it signifies mature castrated male cattle used for draft purposes, though in Continental Europe the term has sometimes been applied to a male not castrated.

HISTORICAL SKETCH OF THE BOVIDÆ.

Domesticated cattle have been derived from wild species of the genus *Bos*, which is one of the largest genera of the family Bovidæ. The members of this family, like all ruminating mammals, possess hoofs with an even number of toes. Among the noticeable features which separate them from other ruminants are the persistent horns with a bony horncore.

The earliest traces of hoofed animals are fragments of bones discovered in New Mexico, which were found embedded in deposits formed in the geological period known as the Eocene. It is difficult to distinguish the forerunners of these herbivorous animals with hoofs from the carnivorous species that had claws. Both walked on the soles of their feet, provided with digits that might answer either for claws or for hoofs. Before the close of the Eocene period typical hoofs had developed in animals living in both North America and Europe. As hoofs developed some of the digits were lost. In later Tertiary formations many changes of the skeleton took place, which led to the inference that small marsh and forest dwelling animals feeding on succulent vegetation had gradually changed into hard-hoofed quadrupeds fitted for life on grassy plains and provided with powerful grinding teeth capable of masticating coarse and dry herbage. A thickening of the brain case, often bearing horns, the disappearance of the incisor and canine teeth from the upper jaw, an increasing height of the molar crowns, and a reduction of digits from five to two, were some of the important skeletal modifications which may be correlated with the incoming of grasses as a dominant feature of the landscape (Woodward, 1898).

The even-toed Ungulata (Artiodactyla) had their chief development in Europe, and after the Eocene period there was a great variety of forms. The ruminating Artiodactyla became vastly expanded and diversified. They were plastic, adaptable, and eventually migrated to all the continents except Australia, but perhaps did not reach North America until the Miocene period, at which time they were far advanced toward the specialized forms found in the present era. In the Miocene of Europe and southern India remains have been found of deerlike antelopes that were formerly thought to be the lineal descendants of earlier forms of the Tragulidæ. Some paleontologists, however, do not think that the Tragulidæ are in a direct line of bovide ancestry.[1] The Miocene antelopes referred to above, in the opinion of some paleontologists, mark the first appearance of the Bovidæ, which includes sheep, goats, musk oxen, and antelopes, as well as cattle. It is the youngest and most specialized family of hoofed animals, having reached their best development only during the present geological period. Members of the family have been found in all parts of the globe except South America and Australia. In North America it is represented by the bison, musk ox, mountain sheep, mountain goat, and a few allied fossil forms. Remains found in Alaska indicate that some of these species may have migrated from Asia in comparatively recent times. Africa appears to be the center of distribution, although their original home may have been in Asia. The Cervidæ, or deer family, is closely allied to the Bovidæ, but they have solid horns, which branch and are shed annually, while the Bovidæ have persistent, unbranched horns, with a body horn core which is surrounded by a horn sheath that grows continuously from the base. *Antilocapra*, an antelope of western North America, known as the pronghorn, is a connecting link between the two families. Its horns are branched and the horn sheath, but not the horn core, is deciduous. Other characteristics of the family Bovidæ are described by Lydekker as follows:

No members of the family, either living or extinct, possess upper canine teeth, or tusks, which are frequently so strongly developed in the deer tribe (especially when antlers are wanting); and in this respect the hollow-horned are clearly more specialized than the antlered ruminants. Very rarely do they show those tufts and glands on the lower part of the hind legs which form such a characteristic feature in many of the deer.

Further evidence of the specialization or high grade of the family is afforded by the fact that the lower ends of the metacarpal and metatarsal bones, which persist in so many of the deer, have invariably disappeared. Then, again, the lateral toes are very generally represented merely by the lateral hoofs, although in certain cases some small nodules of bone within them represent the skeleton

---

[1] For a discussion of this subject see Gregory.

of these portions of the limbs. Moreover, in some members of the family (although in none of those described here) even the lateral hoofs themselves have disappeared and the main hoofs alone remain.

The geological position of the Bovidæ is shown in figure 8.

THE GENUS BOS AND ITS FIVE SUBGENERA OR GROUPS.

The genus *Bos* is the most specialized division of the family Bovidæ, as is shown by the structure of the teeth and by its late appearance, geologically speaking. Lydekker has enlarged the genus so that it includes the species formerly known under the genera of

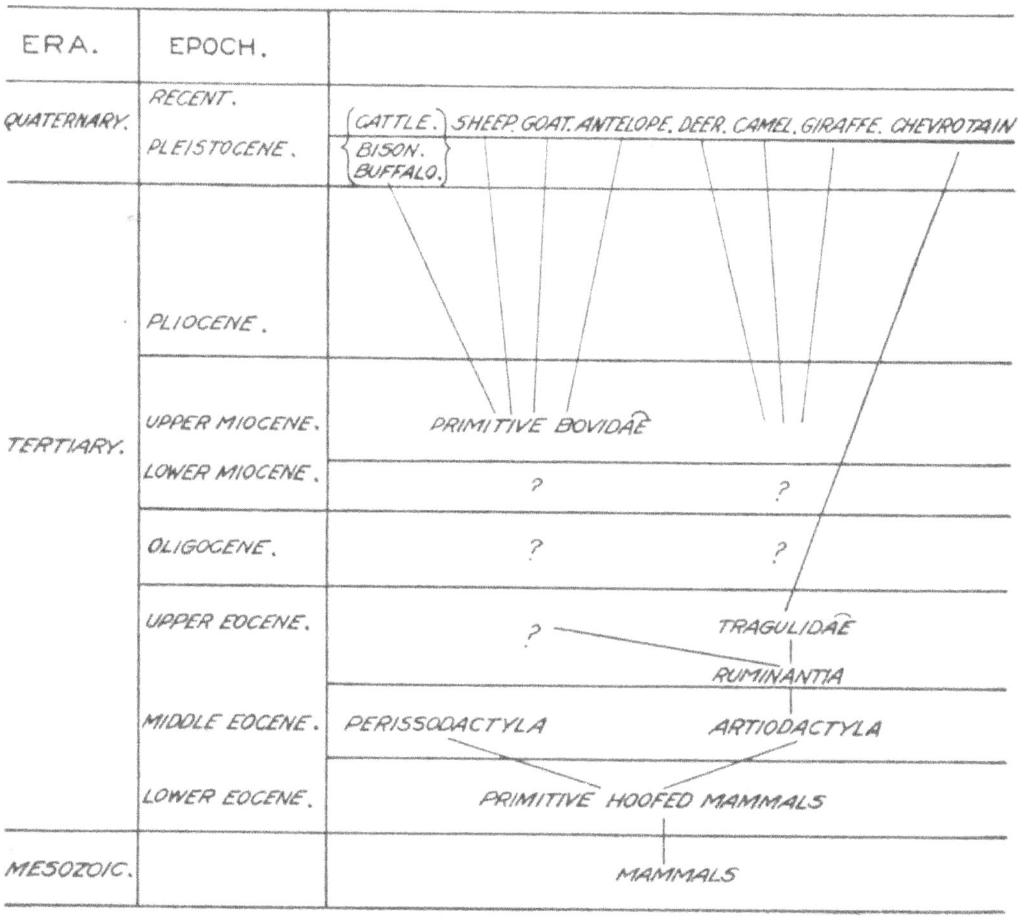

ERA.	EPOCH.		
QUATERNARY.	RECENT. PLEISTOCENE.	(CATTLE. BISON. BUFFALO.)	SHEEP. GOAT. ANTELOPE. DEER. CAMEL. GIRAFFE. CHEVROTAIN
TERTIARY.	PLIOCENE.		
	UPPER MIOCENE.	PRIMITIVE BOVIDÆ	
	LOWER MIOCENE.	?	?
	OLIGOCENE.	?	?
	UPPER EOCENE.	?	TRAGULIDÆ
			RUMINANTIA
	MIDDLE EOCENE.	PERISSODACTYLA	ARTIODACTYLA
	LOWER EOCENE.	PRIMITIVE HOOFED MAMMALS	
MESOZOIC.		MAMMALS	

FIG. 8.—Diagram showing geological position of the Bovidæ.

Bos, Bibos, Leptobos, Amphibos, Bison, Bubalus, Probubalus, and *Buffelus.* He makes five subgenera, however, which correspond closely to the genera of the older classification. The five subgenera or groups he designates as (1) Taurine, which includes our common oxen and the humped cattle of Africa and Asia; (2) Bibovine, composed of the gaur, gayal, and banting; (3) Leptobovine (extinct species only); (4) Bisontine, which includes the yak and bisons; and (5) Bubaline, or buffalo group.

The size of the wild species that are members of the genus *Bos* range from that of the anoa, which is only 3 feet 3 inches in height at the shoulder, to the banting and the gaur, which measure nearly 6 feet in height. Among domesticated cattle we find that some individuals of the Dexter-Kerry, Brittany, and Permian breeds, as well as cattle at the North Cape, are only a little over 3 feet in height. The domesticated water buffalo is sometimes 6½ feet high, and some specimens of the sacred oxen of Ceylon are said to be only 2 feet 2 inches in height.

TAURINE GROUP—SUBGENUS BOS.

The Taurine group is differentiated from the other groups as follows: Typically, the horns are nearly or quite cylindrical, and are situated far apart on a ridge which forms the extreme vertex of the skull that overhangs the proper occipital surface of the latter; the forehead of the skull is flat, elongated, with a long interval between the bases of the horn cores and the sockets of the eyes, which are not tubular; the nasal bones are relatively elongated; the back alone is nearly straight, except in the zebu; the hair is uniformly short; the legs are typically without sharply defined "white stockings;" the seventh or last cervical vertebra is short; the spines of the dorsal vertebræ are of moderate height and slope regularly away to the lumbar vertebræ, thus producing the comparatively straight line of the back. The upward production of the vertex of the skull, so as completely to shut out the occipital surface in a front view, and the abbreviation of the parietal zone, indicate that the Taurine, Bibovine, and Leptobovine groups are the more specialized of all the oxen, but as regards the vertebræ the Bisontine group is more advanced than the Taurine. (Lydekker, 1898.)

Members of the Taurine group formerly ranged from Europe, Asia, and North Africa, although none are found wild to-day except where they may have escaped from domestication. All domesticated forms without the hump Lydekker reduces to one species, *Bos taurus typicus;* those possessing the hump, to *Bos taurus indicus*. These two types have played the greatest rôle in civilization of any of the Bovidæ and have no near wild representatives now living.

BIBOVINE GROUP—SUBGENUS BIBOS.

All of the Bibovine group are humped forms and are natives of southern India. The forehead is shorter than that of the Taurine group, the width at the base of the horns is less, the tail is relatively shorter, and the legs are wide from hock to hoof. The banting and the gayal have been considered by some zoologists as distinct species, while others regard them only as forms of the gaur, one of the largest and most magnificent members of the family.

LEPTOBOVINE GROUP—SUBGENUS LEPTOBOS.

This group consists of extinct species only, and the different members will be treated later.

BISONTINE GROUP—SUBGENUS BISON.

The important members of this group are the European bison, the American bison, and the yak. They may at different times have been crossed with *Bos primigenius* and *Bos longifrons*, as the American bison is being used at the present time in the United States with some of the best beef breeds with the expectation of producing a better breed of beef cattle than any now known. The resulting cross is commonly spoken of as a "cattalo."

Bos grunniens, or the yak, ranges over nearly the entire central part of Asia. The domesticated yak, though somewhat smaller in size, is probably derived directly from the wild form. Kohler, however, thinks the domesticated form is a cross between the wild yak bull and a domesticated cow of the Taurine group. Regel (1884) also is of the opinion that there is yak blood in the long-haired cattle of the upper Oxus.

BUBALINE GROUP—SUBGENUS BUBALUS.

The buffalo is a domesticated animal of considerable importance in southeastern Europe and southern and eastern Asia. It is used as a draft animal and for beef production, and in some sections is the principal dairy animal.

The domesticated buffalo was known in Europe previous to Roman times. It was first introduced as a domesticated animal into Italy at the end of the sixth century. Before that time it was common in the region of the Danube and had probably come from Asia. There was, however, a Pleistocene form, *Bos antiquus*, the Algerian buffalo, which roamed from Algeria to South Africa, and it is possible that the different species of domesticated buffaloes have originated from several wild species of this group.

EXTINCT SPECIES IN RELATION TO ANCESTRY OF DOMESTICATED CATTLE.

In considering the extinct species which may possibly be ancestors of domesticated European varieties of cattle one can not at present go back further in geological history than the horizons in the Pliocene epoch.

THE BANTING. (FROM LYDEKKER.)

PLIOCENE EPOCH.

Geological distribution of the principal species of wild oxen.

Geological epoch.	Asia.	Africa.	Europe.	America.
Recent.	Bos primigenius. namadicus. indicus (zebu). Bibos gaurus (gaur). frontalis (gayal). sondaicus (banting). Peophagus grunniens (yak). Bubalus (many species).	Bos indicus. Bubalus caffre (5 races).	Bos primigenius. Bison bonasus. caucasicus.	Bison bison.
Pleistocene.	Bos primigenius. namadicus. Leptobos fraseri. Bison priscus. Bubalus palæindicus.	Bos primigenius. mauritanicus. Bubalus antiquus. Bison priscus.	Bos primigenius. Bison priscus. Bubalus pallasi.	Bison latifrons. occidentalis. antiquus. crassicornis. alleni. ferox. bison.
Pliocene.	Bos acutifrons. planifrons. Leptobos falconeri. Bison sivalensis. Bubalus triquetricornis. acuticornis. platyceros.		Leptobos elatus. (etruscus.)	Bison [?].

The Leptobovine group in Pliocene time is represented by at least two extinct species—*Bos elatus* and *Bos falconeri*. *Bos elatus*, the Etruscan ox (fig. 9), lived in France and Italy in the late Pliocene. The horn cores of the male grew outward, then curved gradually upward, with an inward tendency at the tips. The limb bones indicate a comparatively slightly built animal. The lower molar teeth have a small additional column on the inner side. Depéret considers *Bos elatus* a bison because of its dentition. *Bos etruscus* (Falconer), originally described as a separate species, is, according to Dr. Forsyth-Major, merely the female of *Bos elatus*. *Bos falconeri* is imperfectly known, but is apparently distinguished from *Bos elatus* by the more slender form of the skull of the male and the more upright direction of the horn cores, of which the bases alone are preserved. Remains of this species are found in the fresh-water deposits of the Siwalik Hills of India, laid down in the early Pliocene.

These species are closely allied to the banting (Pl. XIII), as is shown by the curvature of the cylindrical horns, the shape of the nasal bones, and in the shortness of the skull. On the other hand, the horn cores of the bulls are situated far below the vertex of the skull, midway between the occiput and the orbits. The cows are hornless. Probably the banting (*Bos sondaicus*) is their nearest modern relative, which Keller says can not be the ancestor of the large European breeds because the occipital parts are too prominent.

Turning to the Taurine group, we find that *Bos acutifrons*, the Siwalik ox (fig. 10), is found in late Pliocene deposits. It was a large animal, with angulated frontals and with enormous horns, measuring about 10 feet from tip to tip. Rütimeyer regarded this and *Bos planifrons* as forms of *Bos primigenius*, while Lydekker considers them as distinct species. *Bos planifrons*, with shorter horns and flattened frontals, may have been the female of *acutifrons*.

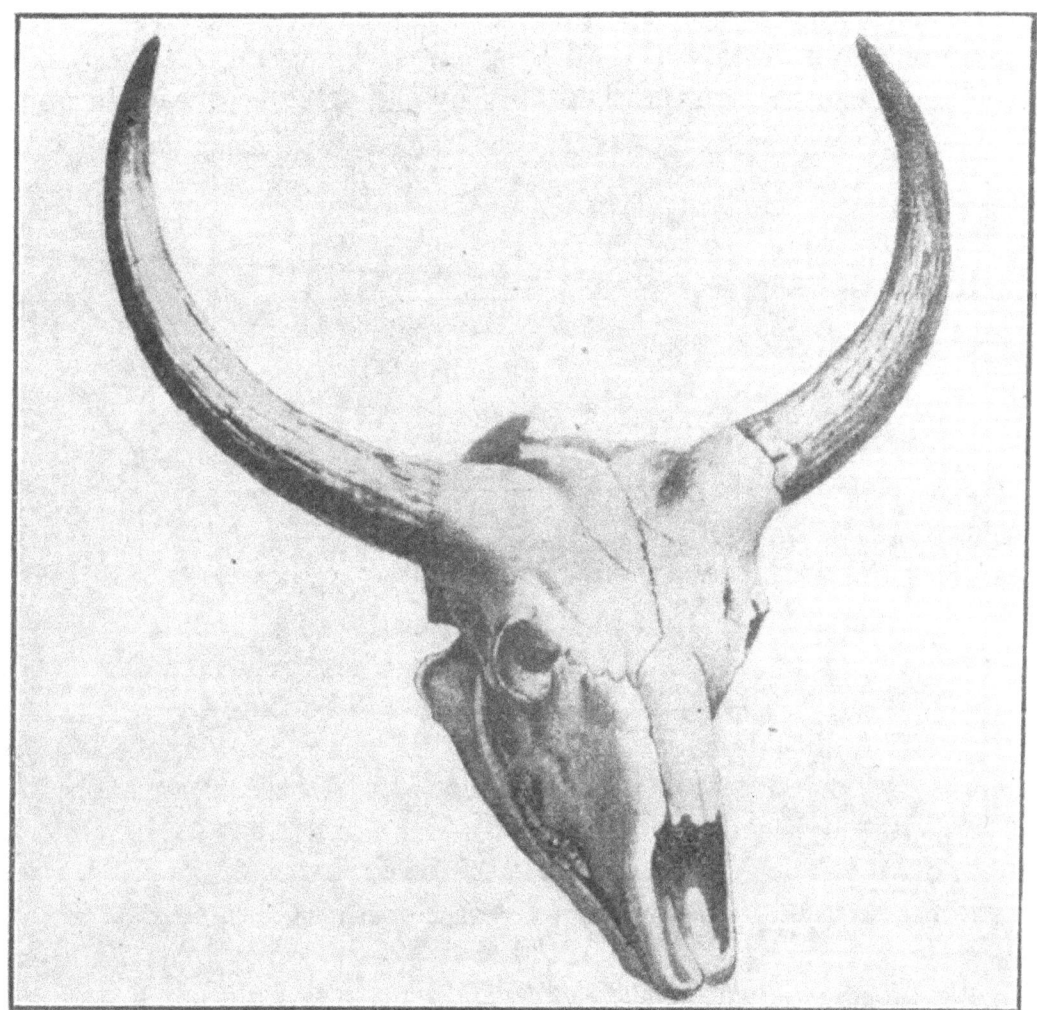

FIG. 9.—Skull of *Bos elatus*. (From Rütimeyer.)

Ewart (1911) has found a modern type of *acutifrons* among the skulls in the Roman fort at Newstead, England.

PLEISTOCENE EPOCH.

In this epoch the species of most interest in the Bubaline group is the Algerian buffalo, previously referred to. (See fig. 10.) A prominent member of the Bisontine group was *Bison priscus*, which roamed over Europe, Asia, and North Africa. This species is probably the ancestor of both the European and American bison. The

Leptobovine group apparently contained only one species, the last member of the group, *Bos fraseri*, and even this is imperfectly known, as only few fossils have been found in the Narbada Valley, India. It resembles *Bos falconeri* in most respects, but also must be closely allied to the banting, because the horn cores are situated some distance below the vertex of the skull.

In the Pleistocene epoch the Taurine group is represented by *Bos taurus mauritanicus* and *Bos namadicus*. *Bos taurus mauritanicus*, so named by Thomas, is probably identical with *Bos opisthonomus* of Pomel, that lived in Algeria and Tunis until historic times and may be only a variety of the European wild ox, *Bos primigenius*, from which it can be distinguished only by a shorter forehead, larger and

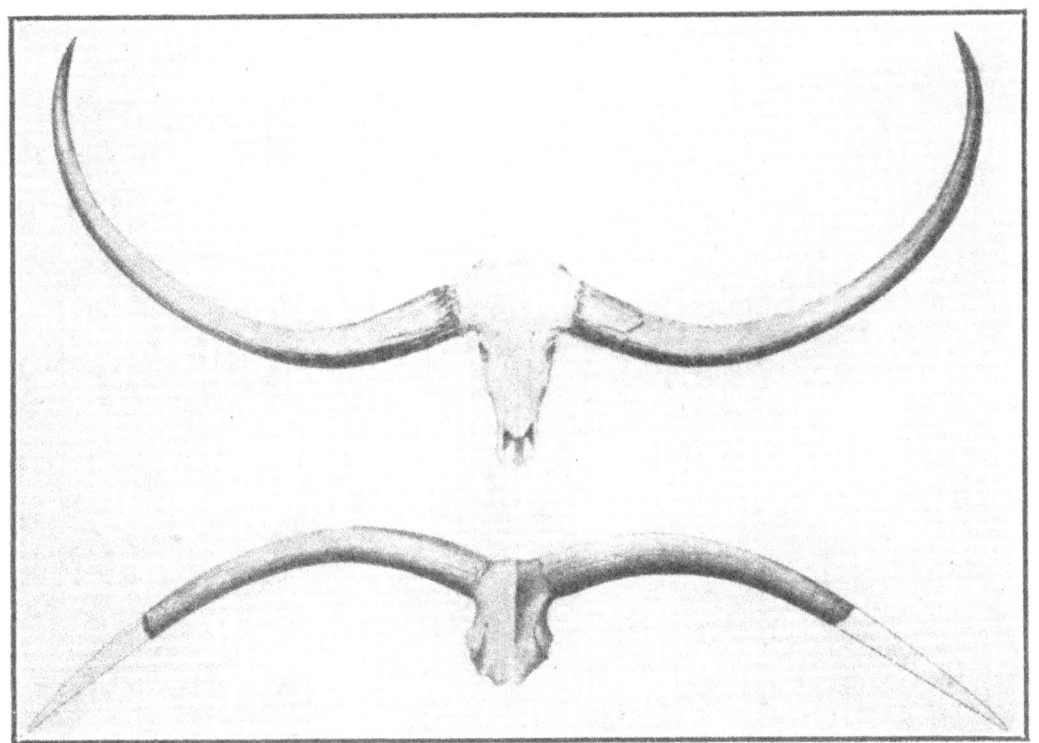

FIG. 10.—Skulls of Algerian buffalo (upper figure) and *Bos acutifrons* (lower figure). (From Lydekker.)

more slender limbs, and with horn cores which curved less forward but more downward. *Bos namadicus*, the Narbada ox (fig. 11), first described by Falconer and later called *paleogarus* by Rütimeyer, is one of the best known species of extinct Indian oxen. It is so closely allied to *Bos primigenius* that it is now considered as the Asiatic and probably older form of *primigenius*. In some specimens the horn cores are somewhat flattened at the base, which shows a close relationship to the Bibovine type. Lydekker suggests that it may have been the ancestor of both the Bibovine and Taurine types, and at the same time a descendant of *Bos acutifrons*. *Bos namadicus* was a contemporary of early man in India during the Old Stone period and has recently been found in the lowest layers of the deposits at

Anau, Turkestan (Dürst). The cranium of *Bos namadicus* differs from that of *primigenius* in the following points: The short premaxillæ, which do not reach the nasals; the low position of the occipital crest relatively to the horn cores; the arcuated intercornual ridge; the intrusion of the temporal fossæ on to the occiput; the concave plane of the latter, and the regular curve of the occipital crest. In almost all of these points in which the cranium of *Bos namadicus* differs from that of *Bos primigenius* it approaches to the crania of the genus *Bibos*. The peculiar forward curve of the horn cores of this species, as is well shown in the profile view (fig. 11), presents considerable resemblance to the curve of the horn cores of the yak.

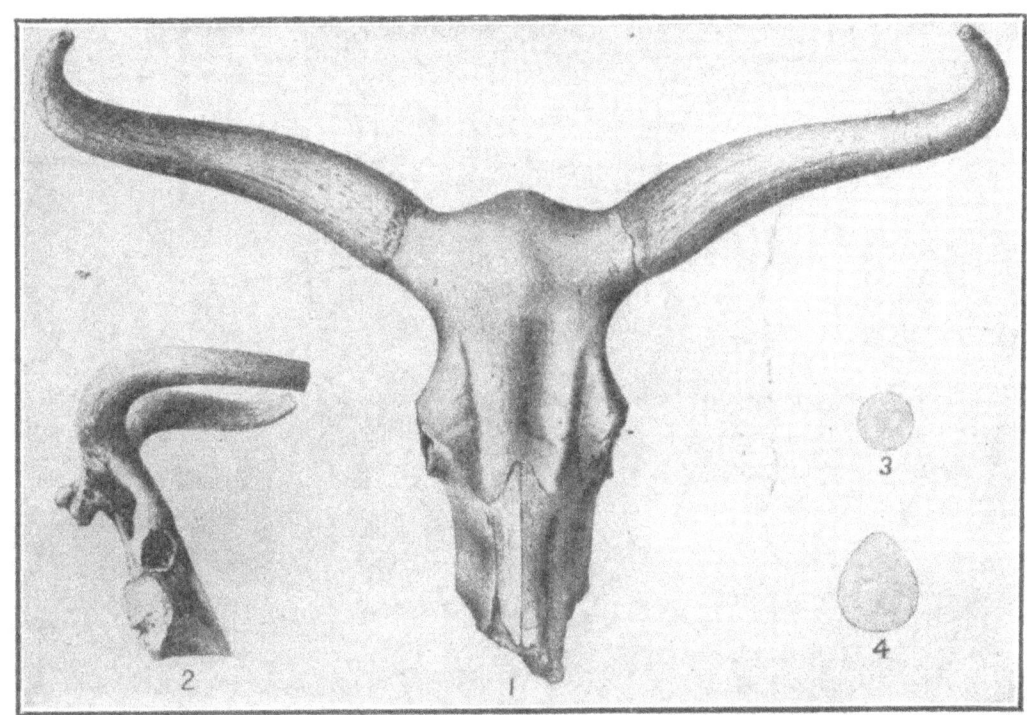

FIG. 11.—Skull of *Bos namadicus*. 1, Frontal aspect; 2, lateral aspect; 3, cross section of horn near tip; 4, cross section of horn near base. (From Lydekker.)

This, however, can not be taken as of any importance in showing kinship between the two animals, as the forms of the crania of the two are so entirely different (Lydekker).

All species of *Bos* which lived in the Pliocene and Pleistocene are now extinct, although the blood of several forms of *Bos primigenius*, *Bos namadicus*, and *Bos priscus* may still flow in the veins of our domesticated cattle. To these species and the representatives of the genus during the Recent or Alluvial periods we must look for the genealogy of our cattle. On the other hand, probably all of the species of the Recent period have played a part in the history of cattle raising.

RECENT PERIOD.

Bos primigenius.

Bos primigenius (Boj, 1827), a contemporary and probably a variety of *B. namadicus*, was a large and stately animal, being 6 or 7 feet high at the withers. It roamed over western Asia, northern Africa, and the entire continent of Europe during the Pleistocene and Recent periods. Like its near relative the European bison (*Bos bonasus*), it was a forest-loving animal and, judging from old pictures

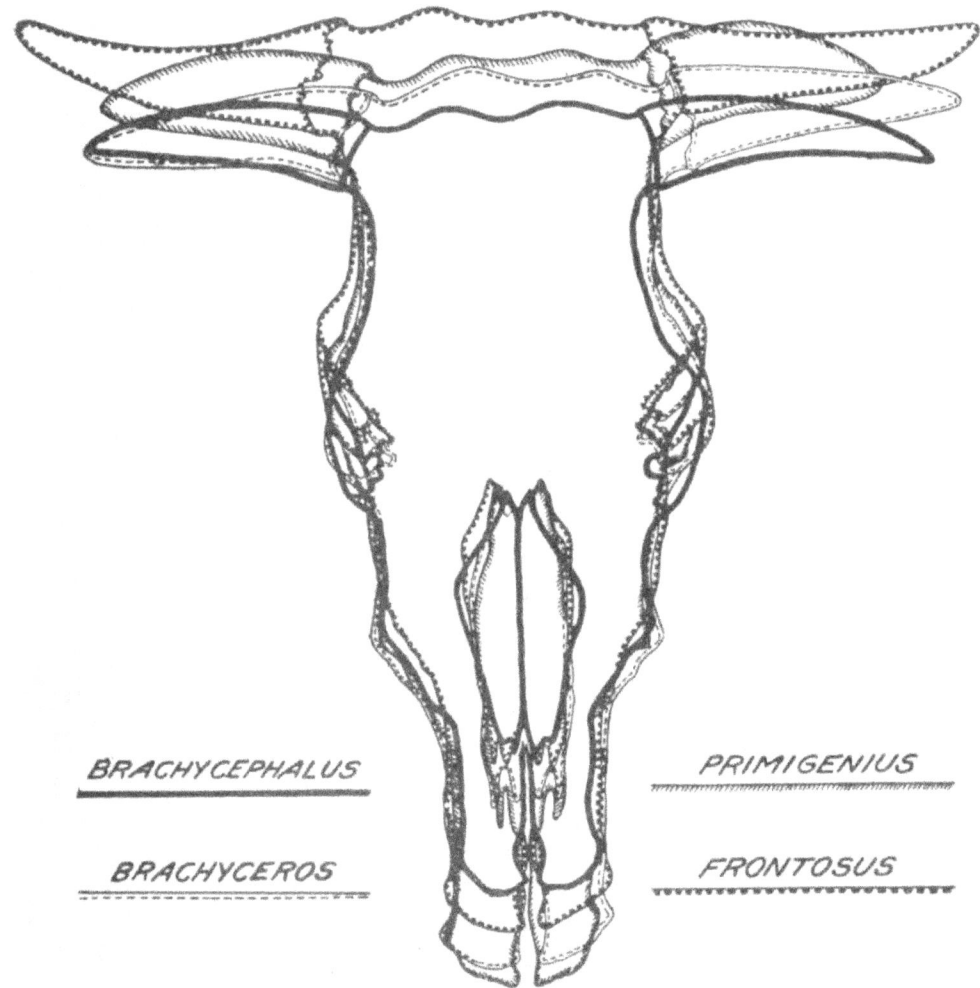

Fig. 12.—Frontal aspect of four types of skulls. (After Wilckens.)

and inscriptions, it had a hairy coat, which varied in color from black or dark brown in summer to gray in winter. A light-colored ring encircled the muzzle, and along the back was a white stripe. Unlike the bison, it had no long hair about the head and neck. To the old Teutons, who used its large horns for drinking vessels, it was known as the aurochs, or ur. But after its extinction the latter name was applied to the bison, which unfortunately has led to much confusion. To the Russians it was known as the tur or thur. Cæsar speaks of it as urus.

That there were large numbers of them is shown by the numerous fossil remains found throughout a wide region. One of the best skeletons ever found is that of a female apparently from 6 to 8 years of age. The skeleton was dug up in 1887 from the bottom of a peat bog at Guhlen, on the shores of Schweiloch Lake, Brandenburg, Germany, and is now in the zoological collection of the Agricultural High School of Berlin.

This skeleton, as described by Nehring (1888), is very similar to that of the cattle of the lowland and steppe breeds found in Europe to-day, except in such changes as would naturally come about by domestication. The horns of *Bos primigenius*, though slender, were long and strong, forming at first a half circle, then extending outward and a little to the rear. At about the middle they begin to turn to the front and end in a point turned a little upward. The forehead and face were long and narrow, with slightly concave surfaces. The whole cranium was somewhat flattened, the contour lines being comparatively straight. Measurements agree relatively with those of the wild cattle of Chillingham Park, England. The forehead was long, narrow, and quite flat. The length of the forehead was 47 per cent of the length of the entire skull. The size of other skeletons of the ur varies from that of our domesticated cattle to over 6 feet in height at the withers and to 12 feet in length.

The first appearance of *Bos primigenius* was in the Pleistocene period, when Europe had a warmer climate than at present. It was a contemporary of *Bos priscus* (the ancestor of the European and American bisons), *Bos bonasus* (European bison), the mammoth, the Irish elk, and other large animals. *Bos priscus* was more numerous before the appearance of *Bos primigenius*, by which it may have been driven out. The remains of *Bos primigenius* are found in all the earlier pileworks of the Lake Dwellers. It was first domesticated in Neolithic times, and later the wild form was driven out by man.

There is much evidence to show that the wild ur or urus has lived within historic times. It is mentioned by Cæsar, who saw it, or knew of it, as an inhabitant of the Hercynian forest. Seneca speaks of both tame and wild cattle. Tacitus and Pliny say that the horns of these cattle, used as drinking horns, sometimes held as much as 2 urs (12 liters). In the Niebelungen Lied, Siegfried kills a wisent (bison) and four ur. In an old chart, made in 1284, the urus is said to exist between the upper Duna, the Dnieper, and the Carpathians, the same region in which he is thought to have become exterminated in the seventeenth century (Beltz, 1896).

Fraas tells of two Roman statuettes of oxen which were dug from a depth of 9 feet in widening a railway in Swabia, in 1895. One represented a bison, the other an aurochs. So it is presumed that both lived in the Black Forest in Roman times; one roamed in the woods of the highlands, the other in the lower meadows.

A painting, presumably made about 1500 and found in 1827 in Augsburg, represents a rough-haired maneless bull, with large head, thick neck, and small dewlap. Its powerful horns turn forward, then outward, and are light colored with black points. The color is sooty black, with a white ring about the mouth. A copy of the picture is in Griffith's "Animal Kingdom," a translation of Cuvier's "Le Regne Animal," Volume IV, and is here reproduced (Pl. XIV, fig. 1). No one knows who painted the picture. Nehring (1896)

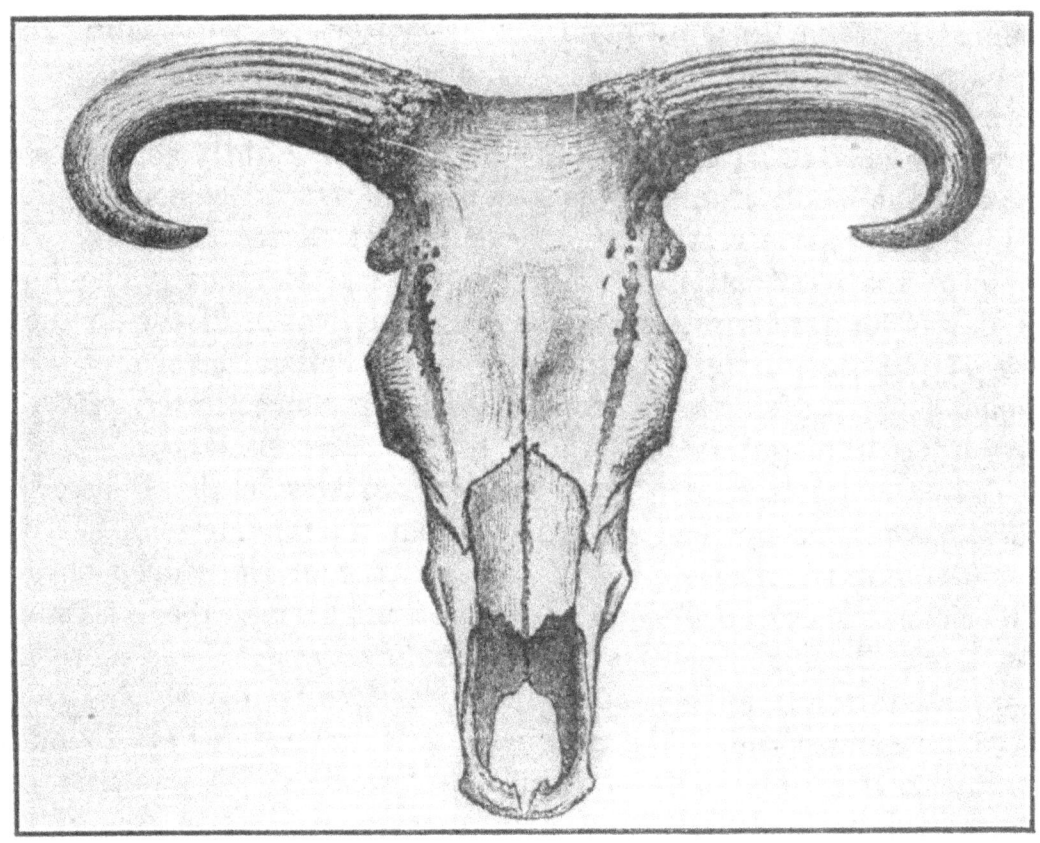

FIG. 13.—Skull of *Bos primigenius*. (From Werner.)

thinks it represents a wild form, while Keller (1897) says it has too fine a nose to be any but a domesticated animal.

In 1889 two golden cups, on which were engraved pictures of cattle, were found in a grave at Vaphio, near Sparta (see Pl. XV). These cups, now in the museum of the Archeological Society at Athens, are evidently the work of a master artist of the Mycenæan period, about 150 B. C. On one is represented a hunting scene with three wild oxen; on the other is a wild ox held by a man, who has fastened a rope about the hind leg of the beast. Two other oxen appear peaceful and domesticated. These figures are supposed to illustrate hunting, capturing, taming, and domesticating. Homer does not mention this wild ox; the Phenician metal worker does not depict him; Egyptians always represent cattle as tamed; so Keller (1897) says the artist must have represented European cattle, and as Europe possessed only two, the bison and the ur, this surely is not the bison and so must have been a representation of the ur.

Krause (1898) is of the opinion that the cups were of Babylonian and not Greek workmanship, because the wild steer is represented in a land of date palms. This opinion is refuted by Keller (1898).

A skull, preserved in the castle of Bromberg, Prussia, shows three spear wounds on the forehead. This is surely some evidence that the urus lived so recently in Europe that many European breeds of cattle may be his immediate descendants, although Pallas (1769), Bojanus (1827), and Jarocki (1830) maintained that no one in historical times had seen a living specimen. This is contrary, however, to the opinion of Gesner, Buffon, Cuvier, and many other zoologists who have studied the problem.

Perhaps the best affirmative evidence that both the urus and the European bison lived within historical times is furnished by Baron Herberstain, who lived during the first half of the sixteenth century. According to his own statements he saw both of these animals when he tarried at the court of King Sigismund August of Poland during a journey to Moscow.

The following is a free translation from the German of Nehring (1896):

> Of the wild animals in lands belonging to Lithuania, besides those native to German soil, is one which they call "suber." It is called "bison" in Latin, while Germans call it "aurochs." Closely related to it is another animal, "tur," or Latin "urus." We Germans call it "bisont" incorrectly, for its form is that of a wild ox. Its color is nearly black, with a grayish stripe along the back.
>
> The suber is considerably different from the wild ox; the head is short and the forehead broad. The horns are far apart, but with the points turned toward each other and are effective as a means of defense. Horns have been found so large that 3 large men could sit between them. They are much thicker and shorter than the horns of the urus. The suber is much higher at the withers than at the rump. The hair is coarse and hard, and not so black as that of the urus. Along the throat and neck the hair is much longer than on other parts of the body.

Under one of Herberstain's pictures is the following statement: "I am the urus which the Polonaise call tur, the Germans aurox, and the vulgar bison." Under another, "I am the bison that the Polonaise call suber, the Germans wysent, and the vulgar urochs." It is this Babel of tongues which has helped to obscure the point at issue. Some of the older German naturalists, including Brehm, have called the European wild ox "aurochs" and the bison "urus." French, Italian, Swiss, Belgian, and English naturalists have called the bison "aurochs" and the wild ox "urus."

The first edition of Herberstain's travels was published in 1549 under the title of "Rerum Moscovitcarum Commentarii," and con-

Fig. 1.—The Augsburg Painting of the Urus.

Fig. 2.—Herberstain's Picture of the Urus. (From Nehring.)

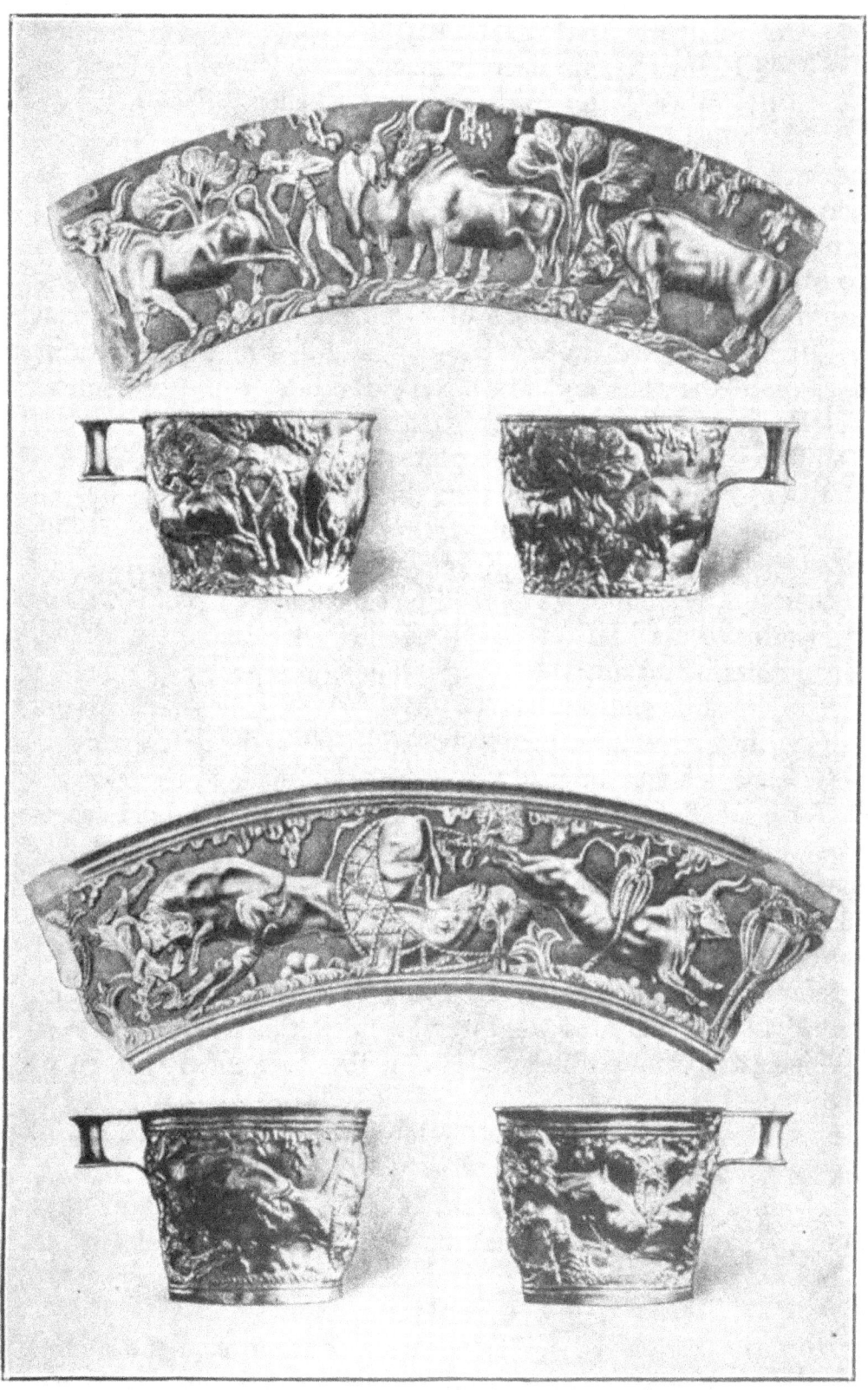

THE VAPHIO CUPS AND THEIR SCROLLS. (FROM RICHARDSON.)

tains no pictures of either urus or bison. The editions of 1551 and 1556 contain pictures of both. An edition of 1557, published by Johannes Steelsius, says that the forest cattle (*Boves sylvestris*) differ from domestic only in being black with a white stripe along the back. An edition of 1557, by Wolfgang Lazius, contains no picture. In 1563 Heinrich Pantaleon translated the work into German under the title "Moskowiter Wonderbare Historien," in which is a woodcut that Wilckens (1885) says is not a representation of the urus but rather that of a castrated domestic ox. The horns appear to be lyraform, while those of fossil *Bos primigenius* are not. Wilckens also thought that Herberstain had never seen an urus himself and went to Moscow but once. Nehring (1896 and 1897), who has carefully studied the various editions of Herberstain's work, says that Herberstain went several times to Moscow and had many opportunities to see both the urus and the bison. Pantaleon, his translator, says that Herberstain had a great regard for truth.

The edition of 1551 contains the plates of bison and urus for the first time, and it expressly states that the plates were added by the author himself, who lived for 10 years after its publication. The pictures were not in the first edition because Herberstain did not possess them until about 1550. He allowed Gesner to publish them in an appendix to his "Historia Animalium" in 1553.

No one knows who made the picture, but it was probably done by an artist in Poland who had seen a urus. As to the shape of the horns shown in the picture, it may be said that artists were rare in those days, and it is not an easy task for the uninitiated to draw the correct form of horns on paper. Besides, the fossil horn cores would not show whether or not the horns were lyraform.

A poem by Caspar Betius, entitled "De Uro et Bisonte," published in 1558, indicates that Herberstain possessed skins, horns, and hoofs of the urus and the bison in 1552 (Nehring, 1897).

Wrzesniowski (1878) has pretty good evidence that these wild cattle lived in the woods of Jaktorowka until the seventeenth century. The last specimen died in 1627 in the Zoological Garden of Count Samoisky.

Other fossil species of *Bos* closely related to *B. primigenius* are *B. trochoceros*, *B. frontosus*, *B. brachyceros*, *B. longifrons*, *B. brachycephalus*, and *B. typicus*. In fact they are so nearly related that some and perhaps all of them may be considered as varieties of *B. primigenius*.

BOS TROCHOCEROS.

Bos trochoceros received its name from Rütimeyer, who found many skulls of females in the pile-works of the Neolithic and Bronze ages. Because most of the specimens appear to be females we may consider that *trochoceros* is a form of *primigenius* domesticated in prehistoric times. The differences between *primigenius* and *trochoceros* are essentially the same as those which appear at the present time between the wild cattle of Chillingham Park and those of Lyme Park in England. In the latter case the change is regressive rather than progressive, as these cattle are presumably reverting to a wild condition from that of semidomestication.

BOS FRONTOSUS.

Bos frontosus Nilsson is a fossil species found in Sweden. (See fig. 14.) The contour of the skull is irregular, the forehead is broad, and

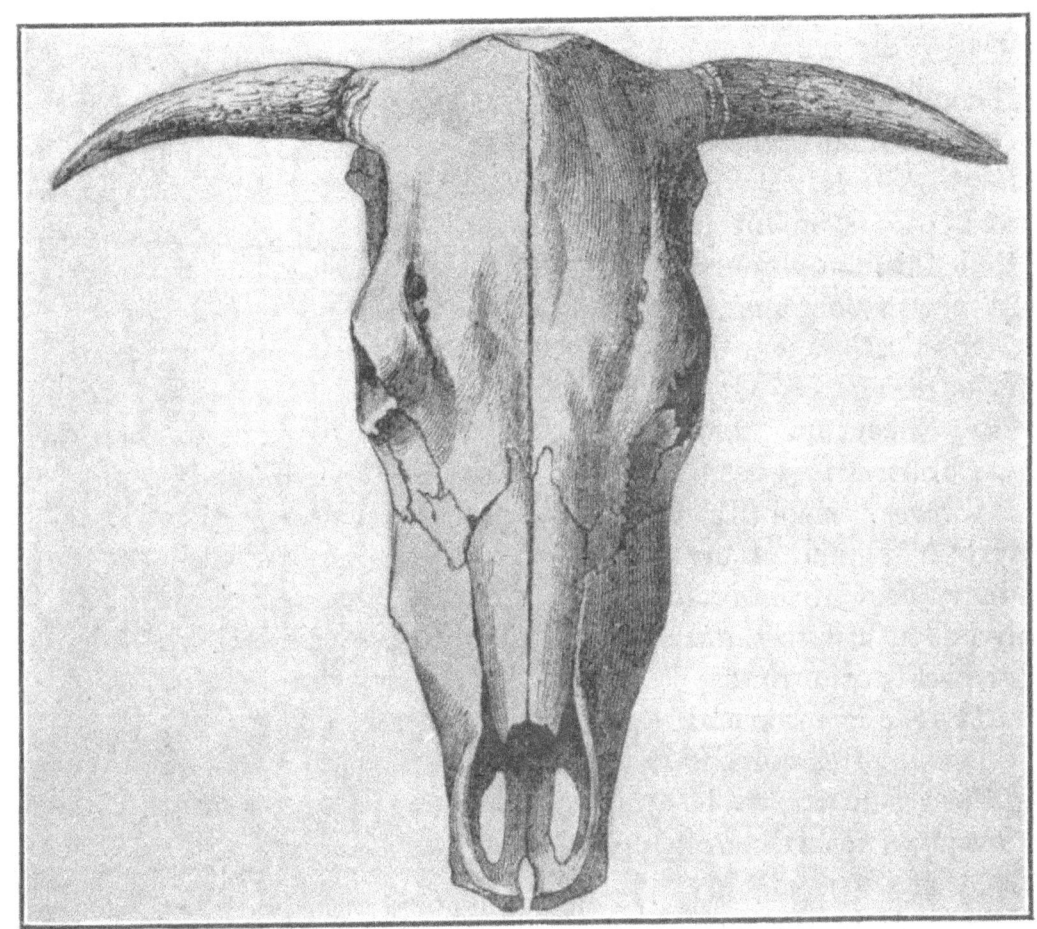

FIG. 14.—Skull of *Bos frontosus*. (From Rütimeyer.)

the horns are situated on a short pedestal. The head is large in comparison with the rest of the body. Nilsson thought it a distinct species having its original home in Germany, and the parent of the mountain cattle of Norway. Rütimeyer (1867) says the Scandinavian remains are those of domesticated cattle. The broadening of the skull is similar to that of the Naita cattle (Darwin) of South America, the Yorkshire swine, and the bulldog. H. von Nathusius says these changes are due to a reduction of muscular activity which accompanies a change in the method of obtaining food. Rütimeyer says this does not explain the shortening and widening of the nasal bones. These changes, however, may be accounted for by the laws of correlation. The differences between a skull of *frontosus* and one of *primigenius* are similar to those between a calf's skull at birth and the skull of the same animal at maturity. Why, then, may not the change to the *frontosus* type have been brought about by an earlier maturity which would naturally follow under the care of man?

Frantzius (1878) thought that *frontosus* originated in Africa, while Arenander says it was a sport from *longifrons* and is the ancestor of *primigenius*. It is more likely that *frontosus* is the product of culture during the polished-stone period, with *primigenius* as the ancestor and not the descendant. Rütimeyer, Studer, David, and others maintain that the bones found in the Stone age at Moosedorf and the Bronze age at Concise, Chevroux, and La Tene are those of *frontosus*, while Keller (1902) and others deny that *frontosus* lived in Switzerland in prehistoric times.

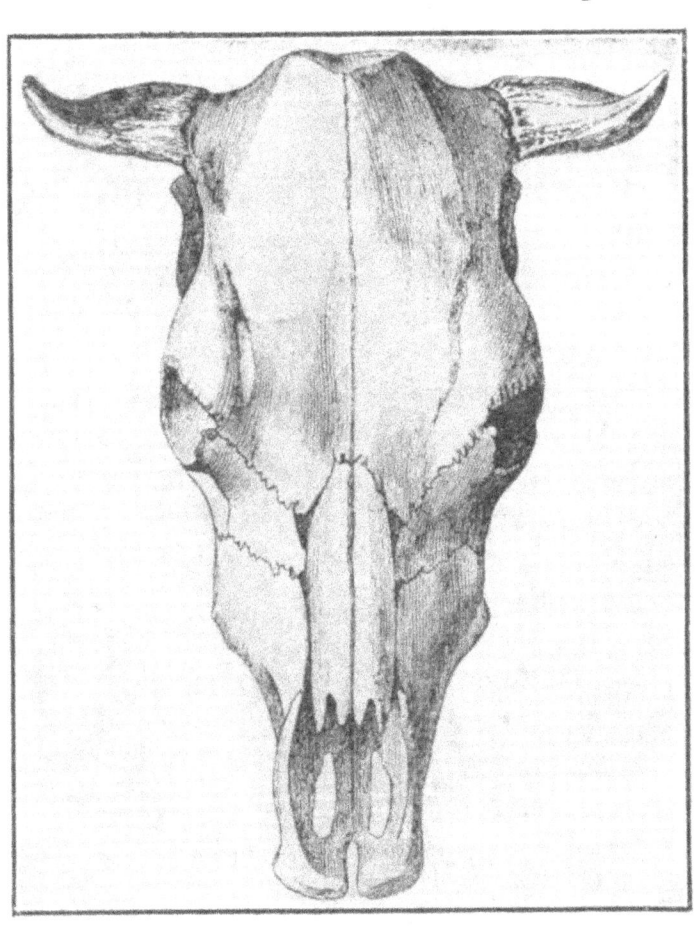

FIG. 15.—Skull of *Bos longifrons*. (From Rütimeyer.)

The Simmenthal and other spotted breeds (Fleckvieh) now in Switzerland and southern Germany are of the *frontosus* type, but Werner says these breeds were carried there by the Burgundians who settled in west Switzerland about 443 A. D., and who brought cattle from their original home in south Sweden. Krämer (1899) is of the same opinion, because neither *trochoceros* nor *frontosus* were in Vindonnissa nor Aquæ Sextiæ during Celtic or Roman times. Although we find a variety of opinions as to the origin of *frontosus*, the reasoning of Rütimeyer, that it was a domesticated form of *primigenius*, seems the most probable.

BOS LONGIFRONS.

Bos longifrons, incorrectly called *brachyceros* by some German zoologists, is another species or subspecies closely related to *primigenius*. (See fig. 15.) *Bos longifrons* is the Celtic Shorthorn found in England and described by Owen. It is probably identical with the species known as the marsh cow, whose remains have been found in the Swiss lake dwellings, and described by Rütimeyer as *Bos brachyceros*. This latter name must be abandoned; Gray in 1837 had also applied it to a species of west African buffalo. Compared with *primigenius*, *longifrons* is much smaller and has a shorter face but a longer and broader forehead. The horns are shorter, and there is a ridge in the center of the poll. It is found with early remains of man's culture in the marshes of Mecklenburg and Harz, in the lake dwellings of the Stone period at Moosedorf, Wangen, Biel, and Wauwyl. It has also been dug from trenches near Bologna, Italy, and was the most important domesticated animal of the Stone age from the North Sea to Italy. But nowhere has it been found wild with certainty. It is represented to-day by breeds in the Alps, in northern Africa, and in Great Britain.

Owen believed *Bos longifrons* to have existed in Pleistocene times, but recent discoveries point to a later origin (Dawkins, 1866). It lived in England during the Roman occupation and is the ancestor of the Welsh and Highland breeds, as the Celts retreated to the mountains with their cattle on the Saxon invasion. In France it is found in the Mousterian period (Mortillet), and was the only bovine species about Lyons during the Gallo-Roman epoch (Cornevin, 1885).

As to the origin of *Bos longifrons*, Rütimeyer, Wilckens, Keller, and Hughes believe it to be a species distinct from *primigenius*. The earliest traces in Europe and the most typical forms of *longifrons* are found on the northern shores of the Mediterranean Sea, the Alpine region, and the Atlantic coast of western Europe. Breeds of cattle in Africa and Switzerland, as well as the zebu of Asia and Africa, possess strong characteristics of this species. Hence it is argued that *longifrons* must have come originally from some Asiatic species, probably *Bos sondaicus*.

Dawkins, Nehring, Werner, and others deny any other than a European origin of *longifrons*. They say that the form of *primigenius* was very variable and that a changing environment and the dwarfing by domestication resulted in the form now known as *longifrons*. Dawkins (1866) says:

A walk into a cattle market will convince the most skeptical of observers that the common ox presents also every variation possible in the shape and direction of the horns. In fine, a very careful comparison of the skulls of *B. urus* in Britain with those of various varieties of *B. taurus*, or the common ox, compels me to believe that there is no difference of specific value between them, those points of difference noticed by Profs. Rütimeyer and Nilsson proving to be peculiar to the individual and not to the species, and therefore useless for classificatory purposes.

Urus could be distinguished from any contemporary taurus by his size, and from the smaller bison by the double curvature of his horns, etc. Nehring (1896, p. 923) cites proof of dwarfing in case of the yak and the banting when kept in the zoological garden. The effect of an unfavorable environment is shown by the condition of the Permian cattle at the present time.

When cattle of two different breeds are placed in a similar environment, similar changes take place, but nevertheless there is always some distinction left. The conformations are never quite alike. The Jerseys, Norman, Angler, and East Friesian cattle live in a similar environment, yet they are different; it may be because they are of different origin. The Jersey cow has good care and abundance of nourishing food, yet she remains small.

Food, domestication, and a change of climate effect great changes, however, and Nehring can see no greater changes between *primigenius* and *longifrons* than between the wild and the domesticated yak. The dwarfing of cattle from a lack of suitable food is well illustrated in northern Russia (Middendorf) and in the Shetland Islands. When Brown Swiss cattle are taken to the steppes of Hungary their horns grow larger, like those of the natives of that region. In the dry year of 1893 young Oldenburg bulls imported to Saxony grew horns similar to those of dry climates and poor food. Pusch mentioned a case which came under his observation the same year. A bull weighed 500 pounds, and showed no signs of growing larger. He came from a cow imported from Pomerania and was of medium size, with some Shorthorn blood. An analysis of the hay showed 0.27 per cent of phosphorus and 0.86 per cent of lime, when normally there should have been 0.43 per cent of phosphorus and 0.95 per cent of lime.

Nehring also calls attention to the fact that the first animals to be domesticated would be young. This in itself would tend to bring about greater changes than if mature animals only were tamed and provided with food. Even if *primigenius* was a large and unruly creature the young might be as easily tamed as if the mature animal was smaller. It is not size, but the disposition to accept the life offered to it, that determines whether or not an animal is capable of being tamed. Yet Nehring (1888) admits that relatives of *primigenius* may have been independently domesticated in Asia and Africa, although he does not go so far as Rütimeyer in thinking *longifrons* one of these independent species, for the following reasons:

(*a*) Color: *Longifrons* is solid in color, black-brown to gray, with a light back stripe, and *primigenius* was solid black with a light back stripe much like the Brown Swiss cattle.

(*b*) Size: *Longifrons* is smaller, but it is because of unfavorable environment, as a raw climate, poor food, in-and-in breeding, and neglect. Dwarfs and small forms of domestic cattle do arise in this way.

(c) Horns and poll ridge: There is a correlation between the two. If the horns are large the poll ridge or crest is flat or even hollow. When the horns are small, the crest is more or less elevated. This is true of the zebu and the yak as well as of cattle. Shorthorned breeds have a crest of medium size when contrasted with longhorned and polled breeds. The size of the horns depends upon food, care, and many other factors.

(d) Skull: A hollow forehead between the eyes and a shortened nose on *longifrons* agrees with modifications which may be brought about by environmental changes. This is illustrated in the Franqueiros and Niata cattle of South America.

Wilckens (1880) says that the lake dwellers would not have tamed *primigenius* when they already possessed the marsh cow; furthermore, that *primigenius* could not have been changed into so small an animal as the marsh cow. The marsh cow, not being found wild, must have been brought there after domestication in Asia. The only alternative is to derive the marsh cow from *Bibos*. The zebu is the nearest to *Bibos* of any of the taurus group and is not found wild except when it has escaped from domestication.

Wilckens divides cattle into dolicocephalic and brachycephalic, and thinks that the long-headed European cattle which form the greater number came from Africa, but as they are not wild there, had wandered thereto from Asia. The short-headed cattle are natives of Europe and descendants of *Bos etruscus*, which, according to Rütimeyer, is the common ancestor of Bibovina.

The recent studies of Ewart (1911) indicate that *longifrons* is more intimately related to the zebu than the wild urus.

In the first edition of "Rinderzucht," published in 1890, Werner (p. 32) says that blood from the Bibovine group can affect only a few breeds in southeastern Europe, a statement which he has omitted from an edition published in 1902.

Adametz (1898) was formerly of the opinion that *Bos longifrons* had its origin outside of Europe, until he saw the fragment of a skull in the museum at Krakow, which, he thinks, throws some light on the subject. The bone in question was found in West Galicia, at a depth of 12 feet, in diluvial strata, as was *primigenius*. It evidently belonged to a mature individual of the female sex, and, according to the rules of Rütimeyer (1862), was of wild stock. Adametz thinks it more like modern European *longifrons* races than *primigenius*. An important point is that the relative distance from the base of one horn to the other is larger in modern *longifrons* races, also in the ancient marsh cow and in his specimen, than in *primigenius*. Adametz concludes that here he has a variety of *primigenius* which probably arose as a spontaneous variation before domestication and is the ancestor of the marsh cow of northern and eastern Europe and of the Polish breeds of to-day, while *Bos longifrons* of the Swiss lake dwellers came from another variation of *primigenius*. He thus finds a European ancestor for one variety of marsh cow and designates it *Bos brachyceros europæus*. Future discoveries may prove Adametz to be right in his conjectures, but the finding of a fragment of one skull is not sufficient evidence to settle beyond all reasonable doubt the question at issue.

BOS TAURUS PRIMIGENIUS *variety* MINOR.

Wollemann (1891) found a skull of a domesticated *Bos*, which was about the size of *longifrons*, but in form resembled *primigenius*, and so gave it the name as given above. There have also been many fragments of bone found in company with the marsh cow among the remains of the lake dwellers which Studer thinks should be placed under *primigenius*.

BOS TAURUS BRACHYCEPHALUS *Wilckens*.

From a study of Alpine breeds in East Tyrol, Wilckens finds a variety of domesticated cattle of which the Duxerthaler breed is a pure type. This type also occurs in the Canton of Wallis. The head is short, the forehead broader than long, and the horn is on a short pedestal. Afterwards he found parts of skeletons in the pile-works of the Laiback moors, which date back to the Old Stone period. To this type he has given the name of *brachycephalus*, or short-headed race.

The bones are quite different from those of the ur, and, furthermore, the remains of ur seldom occur at Laiback. The remains of bison and of these short-headed cattle occur frequently together. The skulls of bison resemble somewhat these short-headed cattle, and the skulls of calves of the modern short-headed breeds bear a still closer resemblance. This is in accord with the view that the individual recapitulates the history of the race, and that ancestral traits may often be seen in the embryo even when absent in the adult. Bison were very abundant in early times in that region, as shown by the numerous remains as well as by the geographical names, such as "Wiesenthal," "Wiesendorf," and "Piesendorf," which are derived from the word "Wisent" (German for "bison"). Therefore, Wilckens thinks that the short-headed cattle of the Alps are of European origin and were brought there by the Celts and crossed with bison, giving rise to this variety, *Bos brachycephalus*.

For these extreme views Wilckens has been attacked from all sides. Rütimeyer thought that *brachycephalus* was not a distinct type of equal value with *primigenius* and *longifrons*, and that the shortheadedness is only the beginning of the pug form already mentioned under *frontosus*. Wilckens admits this influence of culture, but it so happens that many breeds of this type occur in regions where there has been less improvement in breeding than where other types are found.

Keller (1902) and Krämer (1899) say that *brachycephalus* appears first in Italy and is a product of culture from other forms. From Italy it was carried to Switzerland during the Roman occupation. Kaltenegger shows the affinities of the Wallisian breeds to those of the Iberian peninsula and to the old Egyptian representations, and suggests that the origin of the Alpine short-headed cattle is from one of these sources.

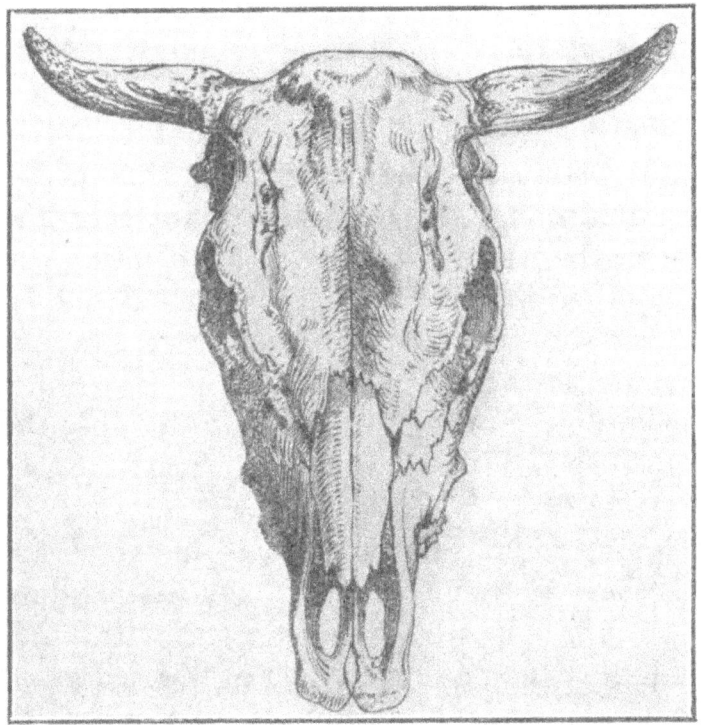

FIG. 16.—Skull of *Bos brachycephalus*. (From Wilckens.)

The measurements of Werner indicate a close relationship between the Iberian and the humped cattle of Africa. The Alpine short heads differ from the humped cattle only in the relative height at the withers and length of horn. Both this excessive length of horn and height at the withers are easily bred off when crossed with other cattle. *Bos primigenius* has been found fossil in Algeria, and it is a reasonable supposition that the humped cattle were crossed with *primigenius* and that the progeny passed over to Spain and thence to Switzerland.

If, however, the fossils found by Wilckens are as old as he believed and are not fragments of *frontosus* (Rütimeyer, 1878) his idea of a distinct short-headed race of ancient lineage is not so easily disposed of. In the meantime we may patiently wait for future discoveries to fill in the vacancy between the paleolithic and modern short-headed kine.

BOS ÆGYPTIACUS AND BOS MACROCEROS.

The first of these names has been proposed by Lydekker to designate the ancient Egyptian cattle as typified by Lortet in the archives of the Lyons museum for 1903, and is a synonym of *Bos africanus* of Fitzinger and Brehm, the question being left open whether these cattle should be regarded as a distinct species or merely as a local race of the domesticated *primigenius*. A modern representative of this type is the ankoli, the longhorned humpless breed of Uganda. Their enormous horns are slender, smooth, upright, and are placed at a much greater distance above the eyes than is the case of the Galla cattle which are a type of humped cattle. *Bos macroceros* is the term proposed by Dürst to designate the type represented by longhorned cattle of Egypt.

Bos indicus.

Bos indicus, or the zebu, includes the humped cattle of Asia and Africa. Col. Yule (1875) says the term zebu is a fantastic name which Buffon heard in a wandering menagerie. The name is not known in India and has probably been derived from the Polish zubr or suber. Perhaps the most notable characteristic of this species is the hump at the withers, although the large drooping ears, the shape of the skull and horns, the white shanks and the grunting cry readily distinguish it from other species. Its habits, such as seldom seeking the shade, and never standing knee-deep in water, are also characteristic, though varying as those of any species must with so wide a geographical distribution. Some races have two humps; the ribs may be 13 or 14 in number, and the horns vary greatly in size and curvature.

No wild form has yet been found, although some instances are known where they have become semiwild and were able to maintain themselves even in a region infested with tigers. Crosses with other breeds of cattle indicate a highly bred animal which has varied much from its original form as it came under the dominion of man, which took place previous to 2000 B. C., if we may judge by carvings and inscriptions in Egyptian monuments that date back as far as the twelfth dynasty.

Hahn is inclined to think that African and Indian zebus had a different origin. No fossil remains have been found in central and southern Africa, so Adametz (1894) says they are of Asiatic origin, with the banting as their ancestor.

It is quite possible that zebu blood has also entered into the longifrons type previously discussed.

Bos triceros.

This is a three-horned humped ox found in Senegal and described by Huet (1891). The third horn is on the nose, like that of a rhinoceros. In all probability this variety is a sport from the zebu.

BOS CHINENSIS.

Swinhoe (1870) describes the small yellow cow of south China as combining the character of *taurus* and *indicus*. Blyth says it is a cross between the two.

POLLED CATTLE.

Calves without horns occasionally appear as spontaneous variations in nearly all districts where cattle are kept. The cause of this variation has not yet been satisfactorily explained. Herodotus (500 B. C.) speaks of the hornless condition of Scythian cattle, in what is now European Russia, as due to extreme cold which dwarfs the growth of horn. Warm, dry climates favor the growth of horns and hoofs. As previously noted, Brown Swiss cattle taken to the steppes of Hungary take on more and more the form of the horns of Hungarian cattle even without crossing (Wilckens, 1876). But if cold is unfavorable to the growth of horn, how can we account for the horns of the musk ox (*Ovibos*), a near relative and an inhabitant of an extremely cold region? In proportion to its size the musk ox has larger horns than most forms of *Bos taurus*.

Whether the polled condition is the result of progression or regression has given rise to much discussion. Dürst says the hornless cattle are the result of domestication. An intermediate stage is the "flaphorn" or movable horn. They are represented in Egyptian works of art. Aristotle knew of cattle with movable horns. Keller has seen them in many places in Africa within recent years.

Flap-horned cattle and sometimes hornless cattle have rough protuberances where the horns otherwise would be, and occasionally small horns may develop as the animal grows older. After many generations these bunches or scurrs disappear. So far experiment has not shown that the hornless condition can be hastened by dehorning. Dürst calls attention to the long horns on the old Egyption representations of the crooked-nosed goat. The modern representatives of this type are either polled or shorthorned. In the time of the Pharaohs, sheep were always horned. All of these instances he cites as a result of domestication.

Arenander (1898) thinks the first cattle were without horns. A study of the geographical distribution of polled cattle in Europe shows them to be more numerous in northern latitudes, which can not be explained by artificial selection by man, for it is unlikely that among people of so many different tendencies the northern people should have always chosen the polled breeds in preference to the horned and the southern people the reverse. We find also among many of the northern people who have polled breeds, that the art of breeding is but little developed. As we go toward the north the proportion of cattle with white hair increases, which evidently is the result of natural selection as in the case of wild animals in a region of snow and ice. Polled cattle are also found in the lake dwellings and hence have been among the early forms domesticated. Again, polled cattle are notoriously prepotent in transmitting this characteristic when crossed with horned cattle. In the struggle for existence polled individuals would be less likely to survive than those with horns, but under domestication would be protected. Thus, Arenander supposes that horned individuals occurred by spontaneous variation and were preserved by natural selection and that polled individuals which occur are reversions.

Arenander (1898) is criticized by Keller (1899), who says that the remains of hornless cattle found in the lake dwellings are in the later ones, while according to Arenander it should be the reverse.

Hehn and Middendorf (1888) think the polled breeds of the north came about because the Scythians wandered farther and farther to the north and took their polled cattle with them. This is stoutly denied by Werner (1902) who thinks the migration of people as well as of cattle was to the south instead of toward the north.

In the Upper Eocene and Lower Miocene the generalized form of deer and antelopes was polled and their associates were tuskless swine and rhinoceroses (Auld, 1887). In the Upper Miocene the antlers of deer were small. Darwin quotes a graduated series of antelopes from the polled condition. The pronghorn as we have seen is an intermediate between the deciduous solid-horned and the non-deciduous hollow-horned forms (Gadow). Hence it may be that the horned condition was not reached until the primitive form of deer, antelope, and cattle had differentiated into forms closely conforming to the present types. That the cows of the oldest member of the cattle kind, *Bos elatus*, were without horns lends support to this view. According to Forsyth-Major, the hornless skulls from the Tertiary deposits of the Val d'Arno in Italy are females of the subgenus *Leptobos* (*Bos elatus* Pomel). Until earlier representatives of the genus are found we must consider the oldest European forms of cattle as having polled females, an indication that in still older species both sexes were without horns.

Ewart (1909), after reviewing the various theories of the occasional appearance of hornless cattle, expresses the view that domestication and the unfavorable conditions which are thought to have reduced *Bos taurus* in size may also have led to the hornless condition. Recent explorations in Turkestan and records from Babylonia lend support to this view, i. e., that the appearance of the polled character is a reversion to the ancestral hornless type.

In a recent paper (Scientific Proceedings, Royal Dublin Society, vol. 12, No. 15, June, 1909), Prof. James Wilson, of the Royal College of Science, Dublin, attempts to show that the hornless breeds of

Great Britain had a common origin, which can be traced back to Scandinavia. This is contrary to the prevailing view that the British hornless breeds have originated independently.

ORIGIN OF THE MODERN BREEDS OF CATTLE.

Regarding the origin of *Bos taurus domesticus* (typicus) and the modern European breeds of cattle it may be asked, Did our modern breeds come from more than one species? If from more than one, can the various breeds be classified according to their ancestry? What was the original home of the wild ancestor or ancestors? Attempts to answer these questions have been made in various ways, and though we are still in doubt and are likely to remain so for some time, we shall review the opinions of those who have endeavored to feel their way through this mist of bewildering evidence.

At one time Cuvier thought our domesticated breeds came from Asia, but later discoveries led him to change his opinion and ascribe to them a European origin, with *Bos primigenius* as the ancestor. Rütimeyer thought that they were all derived from the European *primigenius* except those of longifrons type, whose ancestor must have been of African origin. Darwin derived them from several species, as he conceived *longifrons* and *frontosus* to be species distinct from *primigenius*. Hahn (1896) says they are descended from *primigenius*, with the possibility of some bison blood, as the geographical distribution of the two species is about the same.

Notwithstanding all the evidence which has been produced since Cuvier's time, Middendorf, Nehring, and Werner still hold to his view that the European *primigenius* was the sole ancestor, because *primigenius* remains have been found chiefly in Europe, and Europe, in their opinion, is the home of the domesticated *Bos taurus*.

Frantzius (1878), Pagenstecker (1878), and Keller follow Rütimeyer in thinking Africa the home of at least some of the ancestors of European cattle, while Hartmann is of just the opposite opinion when he looks to the marsh cow for the ancestry of the cattle in the Barbary states. In this he is probably wrong, for they show too close a relationship to the rest of the cattle of Africa.

Keller (1897, 1899) has made a special study of African cattle, and believes in a diphyletic origin of European breeds, the lowland breeds coming from *primigenius* and the *longifrons* breeds from the African zebu. Unlike *primigenius* no fossil remains of wild *longifrons* have been found, hence we should study the migrations of man to seek its original home.

In the plains of Algeria, Tunis, and Morocco the cattle are small, of the short-headed type, resembling zebus in structure and habits, with the exception of the hump. This is known to zootechnicians as the Algerian race. In the more fruitful valleys some of the cattle are larger and possibly have been crossed with European blood. In Nubia the cattle are similar to the Algerian race. Carvings on monuments show that cattle of ancient Egypt likewise were of the same stock. The problem of breeds is complicated in Africa because buffaloes and cattle of other sections have replaced those so frequently lost in epizootics.

The cattle of Abyssinia, known as the "galla," or "sanga," have humps and large horns, but vary much as they have a very wide distribution. The Wahumi or Watussi cattle have horns measuring sometimes 118 centimeters in length and with a capacity of 11 liters (Adametz, 1894). They are also modern representatives of the old Egyptian longhorn of the monuments. Later modifications have given the Bechuana, Transvaal, and Madagascar races. Thus in Africa from south to north there is a constant approach to *longifrons*. The northern branch is shorthorned and humpless, but of ancient lineage, and according to Keller came from Asia in prehistoric times. Its progenitor was a domesticated banting, and collateral relatives are the little marsh cow of the lake dwellers and the Brown Swiss cattle of modern times. Adametz (1898) says an African origin of *Bos longifrons* is impossible. The branches of the Aryan race which have moved the least from this primitive dwelling place (lake dwellings)—the Lithuanians, North Slavs, and Albanians—have cattle to-day which are like those of the lake dwellers. It is probable that these people have the oldest domesticated animals of European origin. The skulls of the marsh cow and those of cattle of some breeds in the Balkan peninsula to-day can hardly be distinguished from one another.

Dürst, from a study of figures and inscriptions on stones, concludes that *Bos longifrons* of the lake dwellers came from Asia in very early times and was domesticated long before Babylonian culture, also that Egyptian breeds came from Asia in prehistoric times. These ancient cattle by their known variability through thousands of years of breeding had three modifications: *Bos macroceros*, which includes the longhorned breeds of Africa, Spain, Portugal, and Brazil; *Bos brachyceros* (*longifrons*), which includes all other horned breeds of Europe; and *Bos akeratos*, the polled breeds, which he thinks may have come from *macroceros* and *brachyceros*.

Kaltenegger (1894), concerning the variations found in the Tyrol, says that the white race, which predominates in the region of Dr. Toldt's brachycephalus division of the people who migrated there, are identical with the white cattle of Italy, southeastern Europe, and western Asia, while the black cattle of the Tyrol are related to the cattle of southwestern Europe and northern Africa. Hence the white cattle of the Tyrol are of Asiatic origin and the black cattle of the same region came originally from Africa.

Summing up the most important of these heterogenous opinions, we find that Cuvier, Werner, Middendorf, and Nehring believe in a monophyletic origin of European cattle, that *Bos primigenius* was the only ancestor and its home in Europe. Adametz and Dürst believe in a diphyletic origin, the ancestors being *primigenius* and *longifrons* and the homes of both in Asia; while Rütimeyer, Frantzius, Pagenstecker, Wilckens, Keller, and Ewart, though believing in a diphyletic or polyphyletic origin, think that at least the home of *longifrons* was in Africa or Asia.

II. EARLY HISTORY OF CATTLE BREEDING, AND CLASSIFICATION OF MODERN BREEDS.

THE CATTLE OF THE ANCIENT CIVILIZATIONS OF ASIA.

Recent explorations in Turkestan have thrown considerable light on the oldest civilization of which we have any record. In deposits of the oldest layers of Anau remains have been found of a wild species of ox, which is undoubtedly *Bos namadicus*, and for the present at least can be considered as the Asiatic form of *Bos primigenius*. The animal was wild, (1) because it was much larger than all the later domesticated bovine animals; (2) because the bones are heavier and harder than those of domesticated bovids; (3) because the bones of other species in the same layer belong undoubtedly to wild animals; and (4) because remains of this large animal are absent in the upper layers of the deposit. In the later deposits, about 8000 B. C., a domesticated longhorned ox appeared, which Dürst regards as a domesticated form of *namadicus*, identical with *Bos taurus macroceros* of Egypt, which was spread at a still later period by tribal migrations. Remains of a small-horned domesticated species occur as a more modern type and possibly may be a stunted form of *macroceros*, unless possibly a smaller bovid may have reached Anau with the other imported domesticated animals. Somewhat later a similar change from the large form to a small form took place in Mesopotamia and also in Europe, but whether it was a dwarfing of the larger species or the introduction of a new species is still an unanswerable question.

The ancient inhabitants of Persia, Babylonia, and Assyria also hunted a wild bovid, *Bubalus palæindicus* Falconer, or a more recent form of that Pleistocene species *Bubalus arnee* Kerr, the Indian buffalo, which is depicted on the cylinder seals of Assyrian kings. Large numbers of these animals were killed by the Assyrian King Ashurnasirpal on the hunting grounds near the Euphrates. Aristotle also mentions the occurrence of the buffalo with horns curved back to the neck in the Persian Province of Kohkand.

Bos primigenius and *Bos priscus* (the Pleistocene bison) are also found fossil together in western Asia as well as in Europe. The bison was wild in Mesopotamia up to Assyrian times. Some teeth of *primigenius* have been found in the bone breccia of Lebanon, which prove it to be coexistent with man, and Dr. Schliemann found the remains of bones of *primigenius* at Troy.

It has been suggested that the unicorn referred to in the Bible down to the time of David may have been *Bos primigenius*, but another alternative is that the unicorn was a straight-horn antelope, which when seen in profile has the appearance of possessing only one horn.

The other domesticated animals of western Asia are much the same as those of Egypt. The zebu was domesticated probably as early as 4000 B. C., and spread from Asia to Africa, so that from very ancient times the distribution was much the same as to-day. Aristotle, Pliny, and Oppian knew of the zebu in Syria, and it may have gradually changed into the steppe breed. (Troltsch, 1902; Keller, 1902.)

THE CATTLE OF EGYPT.

The Egyptians loved their animals, but cattle were the most prized of all. Instead of the lion or the eagle, the bull was the symbol of power and craftiness. The highest goddess was worshiped in the form of a cow. Laborers gave their oxen pet names and conversed with them as we do with dogs. Decked with bright cloths and pretty fringes, cattle were highly esteemed as presents.

Between "ena," the common breed, and "neg," which was rare, we can see but little difference. Hartmann thought there were three breeds and made his division according to the shape of the horns—the lyre-form, the half-moon, and those in which the horns pointed away from each other. Dürst recognizes only the longhorned and the shorthorned, both of which came from Asia and were similar to the cattle of old Babylonia. Polled cattle and flap-horned cattle are there, but they are never represented as at work, so we may suppose them to be "fancy stock." The absence of horns may have been due to spontaneous variations from the other two breeds, but methods of preventing the growth of horns were known to these ancient cattle "jockies" or "fanciers." Sometimes one horn of the bull was bent down by shaving off the horn on the side on which the concavity was desired. This process was accelerated by the application of hot irons.

Dürst says that the humped cattle of Egypt are a variety of the shorthorned, but of a more recent date. They came from Syria, Nubia, and Somaliland, and may be the animal represented on monuments of the fourth dynasty, which, with the exception of the hump, resembles very much modern steppe breeds of eastern Europe.

At the beginning of the fourth dynasty, about 3000 B. C., the Egyptians had domesticated the cow, ass, goat, dog, pig, goose, and antelope (Troltsch). The horse was introduced during the time of the shepherd kings. Sheep, camels, buffaloes, cats, hens, and ducks were later additions. After sheep became common, antelopes and gazelles went out of favor.

CARTHAGINIAN AGRICULTURE.

Notwithstanding that Cicero says the weakness and downfall of Carthage was due to the neglect of arms and agriculture for trade and commerce, there is little doubt that during the height of prosperity agriculture was in good repute, as is witnessed by the writings of Mago and Hamilcar. When Carthage was sacked by the Romans but few books were preserved except those on agriculture. The counsels of Mago for the breaking of oxen to the plow have yet to be improved upon. His writings were the source of much of the information of most Roman writers on agricultural subjects. Most of the cattle of the Carthaginians were derived from the breeds of Egypt.

CATTLE OF PREHISTORIC EUROPE.

THE OLD STONE AGE.

The length of the European Palæolithic period, or stage of civilization in which cut-stone implements were used, can not be estimated with any accuracy. It occurred just before the last glacial epoch, and Mortillet estimates that it lasted many thousand years, while Baranski (1896) says it extended only from 4000 to 1000 B. C., and Müller (1897) thinks it ended in Denmark about 700 B. C. Undoubtedly it ended at a much later period in northern than in southern Europe. Man at that time was a hunter and fisherman, and had no domesticated animals until toward the close of the epoch. The Indo-Germanic invasion occurred, if at all, at about this time.

The Palæolithic shell heaps along the coast of the Cattegat contain bones of *Bos primigenius* and the European bison, which lived there wild at that time, at least 3000 B. C. There are some traces of a smaller ox, but no domesticated ox for certainty.

THE NEW STONE AGE.

At the beginning of the Neolithic or polished-stone period the change from savagery to barbarism was made. During this period it is the common belief that there was an invasion of Europe by people from Asia, who brought with them a few domesticated animals.

Some of the pileworks in the Swiss lakes were erected at this time. Beautiful weapons, household utensils, and ornaments were made of flint and polished stones.

Traces of a domesticated dog are found, and before the close of the period probably the reindeer had been partially domesticated. In the pileworks of Switzerland are found bones of *Bos longifrons*.

In Denmark the Neolithic period extended from about 700 B. C. to 200 B. C. (Müller, 1897.) In Switzerland it began much earlier—extending from about 4000 to 2000 B. C., according to Troltsch. During this period we find cattle, sheep, goats, swine, and perhaps the horse, domesticated throughout northern and central Europe. Possibly some of the later shell heaps were formed about this time, but the most interesting and valuable of our data are from the pileworks about the lakes of Switzerland. Remains of these pileworks were often seen during low water in winter, though little notice was taken of them until Caspar Löhle, a peasant of Wangen, began to collect implements about these works in 1810. The exceptionally low water of the winter of 1853–54 exposed so much of the remains to view that Ferdinand Keller was led to make a careful study of these ancient lake dwellings. So far as our evidence now extends the oldest of these works were built at about the beginning of the new stone period either by the Aryans, who came from Asia, or by people who had borrowed their culture. The similarity of words, the domesticated animals, and cultivated plants, and the distribution of the pileworks, some of which extend as far east as Turkestan, lead to the inference that there may have been a migration. Until future anatomical studies of the different peoples decide the question we must remain uncertain as to whether the culture was borrowed or the people had migrated.

These lake dwellers had a knowledge of agriculture and cattle breeding. They grew wheat, six-rowed barley, millet, flax, caraway, weld, apples, and raspberries. They kept herds of cattle, goats, sheep, and swine. The flesh of the horse was eaten, but it is doubtful if the horse was domesticated at this time. Cattle were used as beasts of burden. Cows were milked and cheese and butter made. A dasher for churning butter was made from a portion of a tree with fasciculate branches, such as is used at the present time in the West Indies.

Bones of *Bos longifrons*, or the "little marsh cow," are very abundant, and this at a period possibly before Babylonian civilization (Dürst).

Bison and *Bos primigenius* are among the wild animals, but there is some evidence that the latter was tamed at this period, giving rise to several varieties. Apparently there were crosses between *Bos primigenius* and *Bos longifrons*.

At Lattrigen, on the southern shore of Bieler Lake, are found what appear to be transitional forms between *Bos trochoceros* and *Bos frontosus*. Troltsch (1902) regards the domesticated primigenius forms as steppe cattle which came from Asia to Europe about 3000 B. C.

Rütimeyer (1862) divides the fauna found in the pilework into two periods. In the first, or age of primitive domesticated races, animal food was mostly obtained by hunting and fishing. The domesticated animals were the dog, sheep, goat, and two races of cattle, *Bos primigenius* and *Bos longifrons*. This period corresponds with the Stone ages of antiquarians. Remains at Wangen and Moosedorf, where there are no tame swine, represent the beginning, and those at Concise the end of the period.

The second period is the age of multiple races of the different animals and begins with the use of metals. Among the new races are the frontosus race of cattle, a larger dog, and a large and a small variety of hog. The period gradually changes to the present, which we may call the age of cultivated races.

THE BRONZE AGE.

At about the beginning of the Bronze age man was slowly advancing from barbarism to semicivilization. The number of domesticated plants and animals increased. In central Europe the lake dwellers were at the height of their development. Cattle breeding at this time held an important place in their industrial life. (Müller, 1897.) The cattle were smaller than the typical *Bos longifrons*, which Rütimeyer says may have been reduced in size by inbreeding. In some places, as at Morigen, sheep breeding was largely replacing cattle breeding. At this time the horse was domesticated, and before the end of the Bronze age the fowl from India was introduced (Troltsch). Recent excavations have uncovered works of art in the palace at Knossos, on the island of Crete, in which *Bos primigenius* is depicted as domesticated and used in bullfighting in pre-Mycenæan times. The Mycenæan age of Greece (about 1500 B. C.) was contemporary with the last of the lake dwellings.

ORIGIN OF THE CATTLE OF MODERN EUROPEAN COUNTRIES.

GREECE.

Neither the rocks of Arcadia, the swamps of Laconia and Acarnania, nor the droughts of Attica prevented the Greeks from successful tillage in the fertile spots of their mountainous peninsula. In Thessaly and Messenia the soil was fertile, and wheat, barley, wine, and oil were the chief agricultural products. Most of the pasturage was poor; thus sheep and goats were more numerous than cattle. Yet there were white cattle in Thessaly, hornless cattle in Borysthenes, and a large breed of cattle, improved by Pyrrhus about 300 B. C., in Epirus. Pyrrhus selected breeding stock according to strict rules, and no heifers were allowed to breed until they were 4 years of age. Some of his cows gave 1½ amphorean (40 liters) of milk per day. Cheese, and probably butter, was made by the ancient Greeks. Arrian says that Alexander the Great imported 2,000 or more head of cattle from India (probably they were zebus).

Country life in Greece was dependent to a great degree upon social and political conditions. Constant risk of invasion compelled the people to live in or near the city walls. The rural arts were practiced by slaves and peasants, yet the science of agriculture was known to the educated, for Columella says that 50 agricultural treatises in Greek had been lost before his time.

ITALY.

Though the Indo-Germans led a pastoral life and knew the cereals only in a wild state at the time of the migration, the Greco-Italians were grain cultivators. We infer, then, that agriculture was adopted after leaving their original home and before settling in Italy and Greece. Italy was a land much better adapted for cattle raising than Greece. The Greeks referred to Italy as the land of cattle. Perhaps the word "Italy" was derived from the Greek word meaning "bull," although some philologists ascribe the origin of the word to "Italus," a mythical king in southern Italy who persuaded his people to turn from herding to tilling the soil. From the Roman laws we may infer that wealth first consisted of cattle and the usufruct of the soil (Mommsen).

Among the Romans the draft ox was considered invaluable, and to kill one was as serious a crime as to kill a man. White bulls were often offered as a sacrifice to the gods. Each Roman province had its own breed of cattle. In general, the different breeds, especially those of Lucania, Umbria, and Sabinia, were large and of the brachycephalus type, when it appears for the first time. In Campania and Siguria the cattle were smaller and of the longifrons type. Latium possessed a close-built, good working breed. The Apenine cattle were hardy but less comely. The smaller breeds in the valleys of northern Italy yielded a good flow of milk, which in the spring was considered a good medicine. Many Romans went to the herds of Switzerland for the cure of tuberculosis.

From Columella's description of the points of bulls, cows, and draft oxen, we may conclude that considerable attention was paid to selection of breeding animals. That cattle were bred in large numbers we know from the Punic Wars, when Hannibal captured 2,000 oxen and at one time offered up 300 white bulls as a sacrifice. At another time he escaped from a snare laid by Fabius Maximus by tying torches to oxen at night and driving them up the slope of the mountains. The Romans, thinking that the Carthaginians were escaping, started to head them off, but were met by an array of wild oxen. Hannibal easily escaped through a defile which was then left unguarded.

Cattle breeding in Italy was influenced by breeding stock which was carried from Epirus in Greece to Lucania. Keller (1902) says that these were of primigenius type, whereas Kramer (1899) does not see *primigenius* represented by any Roman artist.

In northern Italy and Switzerland the larger breeds were spoken of as Celtic cattle. Kramer (1899) thinks the Celts did not bring their cattle with them. When Cæsar defeated the Helvetians in the battle of Bibrachte all of their property was destroyed. After that these brachycephalus types, which were the result of breeding and which the Celts did not know, were taken by the Romans into Switzerland. Like the Roman goat, this type of cattle was carried in two directions, one to the valley of Wallis in west Switzerland and the other to the Duxer and Ziller Valleys in the east. In excavating for a railroad from Lyons to Vaugneray in 1885 many bones of cattle were found among the remains of the Roman village of Lugdunsian, which were of the longifrons type (Cornevin, 1885; Mortillet, 1890). In the Celtic deposits of Liggenthal only *longifrons* and *primigenius* are found. Roman cattle were also taken to England, as we shall see later.

GAUL.

In very ancient times, along the continental shores of the English channel (northwestern France and the Netherlands), the people had cattle of the longifrons type. Farther east the people then living between the Danube and the Alps possessed large and strong cattle that must have been of the primigenius type. The color of these cattle was red with white markings. We may consider this breed as the progenitor of the modern breed of Salers, France. In the meantime the original breed had been crossed with *Bos brachycephalus*, and the resulting cross was known as the "Celtic red."

About the middle of the fourth century the Salie Franks entered Gaul and after much fighting settled in the northeastern part, bringing their large cattle with them. Thus the native *longifrons* was supplanted by the westward march of *primigenius*. Before the close of the next century the remainder of Gaul had been conquered, though but few cattle were introduced (Werner).

IBERIA.

Before the Aryan invasion the people of the Iberian peninsula (Spain and Portugal) had a variety of *brachycephalus* cattle. These cattle were also in north Africa, Corsica, Sardinia, and France. Modern representatives are in Spain to-day. They gradually spread to the east and deviated from the common type. According to Werner the Celts brought primigenius breeds to Iberia and crossed them with the native stock, giving rise to the breeds now known as Brittany of France, Kerry of Ireland, the Welsh Black, Devon, Hereford, and Longhorn of Great Britain, the Duxer and Pinzgauer of southern Germany, and the Piedmont breeds of northern Italy. Kramer (1899) thinks that the Celts did not take their cattle with them as they migrated. It is quite certain that the Celtic breeds had little effect in changing the Iberian native stock.

SWITZERLAND.

Werner states that the aboriginal Swiss cattle were black with white spots, now represented by the short-headed Vogesen and the Freibourg black spotted, but that the latter has lost the short-headed character in crossing.

When the Burgundians settled in west Switzerland in 443 A. D., a new breed of the frontosus type was introduced to Switzerland. According to Ptolemy and Pliny the Burgundians previously lived near the Rivers Vistula and Spree. Some later evidence points to south Sweden as their home. In the Icelandic Edda (written probably about the twelfth century) the island of " Bornholm " is called " Borgundarholm."

It is well to note that *frontosus* was first found in south Sweden. A study of this type may furnish some decisive evidence as to the original home of the Burgundians. The cattle in the mountains of Switzerland still retain the longifrons characteristics, and the improved breeds, like the well-known Brown Swiss, are a result of careful selection ever since the pileworkers lived in that vicinity. (For a history of this breed, see Ringholz, 1908.)

RUSSIA.

Herodotus writes of the polled cattle of the Scythians, in what is now southern Russia, about 500 B. C. Werner says that these Scythians had a breed of brown cattle with long horns, and hornless zebus, as well as crosses between the two. Middendorf finds small hornless breeds in the woods of southern Russia which correspond to the Scythian cattle of Herodotus, and also says that the polled primigenius breeds of northern Russia came from this Scythian stock. In Finland and northern Russia the cattle are small but of primigenius type. The Cholmogory breed of northern Russia was a cross of the native cattle with improved breeds from Holland.

GERMANY.

From Tacitus and other classic authors we learn that in the land of the Germans there were many cattle, which were small, though of the primigenius type.

> It abounds in flocks and herds, but in general of a small breed. Even the beef kind are destitute of their usual stateliness and dignity of head; they are, however, numerous, and form the most esteemed, and indeed the only species of wealth. * * *
>
> Homicide is atoned by a certain fine in cattle and sheep. * * *
>
> Their food is simple; wild fruits, fresh evening or coagulated milk. [A note says this is not cheese, although Cæsar says "their diet consists of milk, cheese, and flesh."]

Pliny says:

> It is surprising that the barbarous nations who live on milk should for so many ages have been ignorant, or have rejected the preparation of cheese, especially since they thicken their milk into a pleasant tart substance, and a fat butter; this is the scum of milk of a thicker consistency than what is called whey. It must not be omitted that it has the properties of oil, and is used as an unguent by all the barbarians and by us for children. (XI, 41.)

Cæsar says the Germans were not studious of agriculture. A cross of the *primigenius* of the old Teutons took place with *longifrons*, which is seen in the lowland breeds of Germany to-day. At the beginning of the fifth century the Allemanni went from south Germany to east Switzerland, carrying the red Celtic *brachycephalus* and crossed it with *longifrons*, giving rise to the Algaüer breed. The yellow breed of Oberinthaler may also have arisen in a similar manner after the Gauls entered Rhetia (Werner).

In northwestern Germany the cattle are similar to those of the Netherlands, while in the southern part they are more like the cattle of Switzerland. In the middle part of the empire are a variety of breeds which are little known outside of their native districts, so that any detailed description of them here must be omitted. A few representatives of the Simmenthal breed, a beautiful modern type of *frontosus*, have been brought to the United States.

A detailed description of German breeds may be found in the works of Dr. Hugo Werner and other German authors. (See also "Cattle and Dairy Farming," United States Consular Reports, 2 vols., 1887.)

THE NETHERLANDS.

The oldest inhabitants of Holland of which we have any records are the Friesians, who dwelt on the shores of the North Sea as early as 300 B. C. They were a peaceable, pastoral people and may have originally migrated from central or western Asia. Little is known concerning the characteristics of their cattle, but it is certain that a portion of them were white and that they were of some religious significance. Two hundred years later another German tribe, the Batavians, came down the Rhine from Hesse and settled near the Friesians, where they drained marshy lands and islands, built dikes, and had numerous herds of large, longhorned black cattle of the primigenius type, which in all probability they had brought from their former home. Since that time cattle-keeping has been the chief occupation of these people when not engaged in defending themselves against the onslaughts of the Normans, Jutes, Angles, and other quarrelsome neighbors.

Referring to a more recent period, Motley, the historian, says:

> On that scrap of solid ground rescued by human energy from the ocean were the most fertile pastures in the world; an ox often weighed 2,000 pounds, the cows produced two and three calves at a time, and the sheep four and five lambs. In a single village 4,000 kine were counted. Butter and cheese were exported to the annual value of $1,000,000, salted provisions to an incredible extent. The farmers were industrious, thriving, and independent.

So fertile were the lands in this region that during the twelfth and thirteenth centuries a tide of emigration set in that direction from neighboring lands less fitted for the grazing and rearing of cattle. The first cattle market in that vicinity was at Hoorn, which was established as early as 1311 A. D. Bakker's recent study of the origin of the cattle in Holland leads to the conclusion that the original color was red and that the black color came from Jutland cattle imported in the latter part of the eighteenth century.

As a good cattle land Holstein is of somewhat later date than the Netherlands. Many colonists went from Friesland, Holland, and Westphalia and settled in Holstein, taking cattle along with them (Hengeweld). By far the larger number of cattle in this region are black-and-white in color, large in size, and noted for giving large quantities of milk. Those in Holland may be considered the most typical of the breed. Many subbreeds, some of them red in color, have been derived from the principal type and are designated by some local geographical name. It is a misnomer to call the breed Holstein, which is only a subbreed, or even Holstein-Friesian, a name adopted by the breeders of these cattle in the United States. The native farmers use the term "Dutch" to designate the cattle, which is more appropriate; as also the "Nederlandish," and "Hollandaise," used respectively by the Germans and French.

In the seventeenth and eighteenth centuries large numbers of Netherlands cattle were imported to the British Isles, most of them going to the district of Holderness and the fertile district of the Tees. Peter the Great imported some to Russia and crossed them with natives, and from the cross has resulted the Cholmogorian breed. In the last two decades of the nineteenth century large numbers of Dutch cattle were imported to the United States.

SCANDINAVIA.

From the ancient Sagas we learn that there were two breeds of cattle in Scandinavia. One was a small white or white-spotted, hornless breed living among the mountains in north Sweden; the other was a large black breed similar to the cattle of Jutland and Denmark. At the entrance of the Goths there was another highly prized, large-horned breed, either red or yellow in color, which appears to have been introduced by them. The Vikings were in the habit of taking their cattle with them on shipboard, and the Norwegian settlers in Iceland in 874 brought their cattle along with them (Malet, "Northern Antiquities"). Thorsin, the Icelander who founded a colony in Vinland, carried cattle with him.

Sundbärg, referring to the cattle of Sweden, says:

> The history of the cattle in our country presents a good many vicissitudes. The law of Uppland, A. D. 1296, describes Swedish cattle as being small, hornless, white or whitish gray, often with dark spots. The Alpine breed in northern Sweden is so still, a race we have every reason to consider as being the oldest in the country.

In the sixteenth century King Gustav I imported breeds from Jutland and Holland. Since then many importations have been made from the lowlands. The famous Thelemarken breed of Southern Norway is a direct descendant of *Bos frontosus*.

DENMARK.

In Denmark there are principally two breeds, the black-and-white breed of Jutland, with an origin similar to that of the Holland cattle, and the Red Danish, a breed whose origin is about the same as the red breed in Schleswig. Both of these have gradually evolved from the native breeds which have been in that vicinity since the dawn of history. Many of the English breeds, especially Shorthorns, have been introduced into Denmark, and, according to Rasmussen, have had an important influence in the development of Denmark from a grain-growing to a cattle country, not so much by the infusion of new blood, as by giving the farmers an ideal as to form and teaching them the importance of good feed and care in the rearing of cattle.

FRANCE.

The French eat much less beef than the English, and thus their breeds of cattle have been developed for the dairy and for draft purposes rather than for meat. The largest and most popular general-purpose breed is the Normandy, a descendant of *Bos primigenius*. There are several subbreeds—one of them, the Cotentine, has been bred mainly for milk, and another, the Augeronne, for beef. The heavy-milking Charolais breed, white in color, is presumably a modern type of *Bos frontosus*. The Nivernais, a subbreed, is probably a cross between the English Shorthorn and the Charolais. The Flamande breed in northeastern France, whose origin is similar to that of the Dutch cattle, is divided into a great many subbreeds, some of which are famous for milk production.

The Limousine breed, of Celtic shorthorn type, is one of the best beef breeds of France, although not so large as many others. Breeds of southwestern France resemble the Iberian breeds, and those of the southeastern part are more like the Swiss cattle.

Besides the native breeds there have been many importations of Shorthorns into northern France, as well as some Jerseys and Holsteins. (For detailed descriptions of French breeds see Werner, De Lapparent, and United States Consular Reports.)

GREAT BRITAIN.

Abundant remains of *Bison priscus* and *Bos primigenius* appear in Pleistocene strata in Great Britain. Both continued to live there for a long time. The bison disappeared first, while *primigenius* continued through the Neolithic period and possibly in the mountain fastnesses until within historic times.

But a short time elapsed between the last of the lake dwellers in Switzerland and Cæsar's entrance into Great Britain, which was then in the pastoral stage, but during the Roman occupation the inhabitants began to pay more attention to the cultivation of plants. Cæsar found large herds of domesticated cattle, but they were evidently of the longifrons type, which was abundant during the Bronze age. The remains reveal a small breed about the size of the Irish Kerry. The small horns were sharply curved forward. Excavations show that since that time the native breed was gradually modified and increased in size with an outward and upward curve of horns. Hughes (1894 and 1896) says these changes could not come about by a cross with primigenius breeds.

Sculptures, coins, and mural paintings of Roman cattle are represented with upturned horns much like some Italian breeds of to-day. Other Italian breeds have horns growing outward. Reasoning from these premises, Hughes thinks that cattle were carried from Italy to England during the Roman occupation and crossed with the native *longifrons*. The semiwild cattle now roaming in the parks of Great Britain resemble the Sicilian and ancient Roman breeds. As Roman cattle were also one variety of *longifrons*, it would appear that the early breeds of Britain were exclusively of the longifrons type. The only alternative is the question as to whether or not *longifrons* is itself a stunted form of *primigenius*.

Another modification of British cattle began when longhorned cattle, of primigenius descent, were introduced from Jutland, Friesland, and the Lower Elbe. Grant Allen, in "Anglo-Saxon Britain" (p. 14), says:

> The early English in Sleswick and Friesland had partially reached the agricultural stage of civilization. They tilled little plots of ground in the forest; but they depended more largely for subsistence upon their cattle, and they were also hunters and trappers in the great belts of woodland or marsh which everywhere surrounded their isolated villages. They were acquainted with the use of bronze from the first period of their settlement in Europe.
>
> The wealth of the people consisted mainly in cattle, which fed on the pasture, and pigs turned out to fatten on the acorns of the forest; but a small portion of the soil was plowed and sown, and this portion also was distributed to the villagers for tillage by annual arrangement.

The Saxons probably brought their cattle with them to England, while the Celts retreated with their shorthorned *longifrons* to the mountains of Scotland and Wales. The descendants of these cattle have furnished the foundation stock of modern breeds in those districts. Later introductions from Normandy and northern Germany have modified the breeds in the eastern and southern countries. Hughes cites the Kerry as the modern breed most typical of the old Celtic Shorthorn, the Highland and Welsh breeds of the cross between the Celtic Shorthorn and Roman cattle, and the longhorn breed as the most typical of the result of a cross with the breeds of the lowlands on the Continent.

It is the general opinion at the present time that the white Park cattle are descendants of some domesticated white breed which have become feral. Wilson believes that all hornless breeds of the British Isles can be traced to a Scandinavian origin, but this does not account for the hornless wild cattle nor the hornless skulls found in the Roman fort at Newstead.

Ewart (1911) finds four distinct types of horned oxen at Newstead, namely, *longifrons*, *primigenius*, *acutifrons*, and a type with a convex forehead and rounded poll resembling *namadicus* rather than *primigenius*. There were also flat-polled and round-polled types of hornless oxen.

IRELAND.

The native cattle of Ireland are presumably descendants of the wild forest breed, as they have characteristics resembling the Welsh and Highland cattle. The remains of *Bos longifrons* are also abundant. At Uriconium, which for a long time was the headquarters of the Roman Twentieth Legion, remains of *frontosus* have been found (Blyth, 1864).

Females were more abundant than the males, an indication that they were domesticated. The oldest annals of Ireland refer to horned cattle, but for a long period hornless cattle also have been quite numerous.

CHANNEL ISLANDS.

It has often been stated that the cattle of the four islands of Jersey, Guernsey, Alderney, and Sark originally came from Normandy, yet most naturalists place Normandy cattle in the primigenius group, while the Channel Islands cattle, the Brittanies, and the Kerries are regarded as descendants of *Bos longifrons*. In our opinion this is the most likely supposition, although no doubt there has been some admixture of Norman blood, especially in the Guernsey cattle.

The deerlike form and color of the Jerseys indicates unmistakably longifrons blood. The cattle of the other islands, by their color, length of leg, and larger body, show a closer resemblance to the Norman cattle, which would be likely, as laws restricting importations of cattle have been less stringent than in the island of Jersey, where no foreign breed has been imported for about 125 years.

ORIGIN OF THE PRINCIPAL TYPES OF CATTLE IN AMERICA.

The first cattle in America were brought in 1493 by Columbus on his second voyage. Subsequently many other cattle were brought from Spain to the New World by the colonists who settled in the West Indies. Some of the cattle escaped from captivity and lived in a wild state where there were rich grazing lands in the wilds of the Antilles. From the West India Islands these cattle of Spanish descent were carried to the mainland both north and south of the Isthmus of Panama. About 1525 some were taken to Vera Cruz, Mexico, where they rapidly multiplied and gave rise to the stock which later became known to the breeders in the United States as "Texas" cattle, and hence are of Spanish origin as well as those of South America.

The Portuguese made settlements at Cape Breton Island and other places, as an adjunct to the fishing industry, about 1525, or even earlier. Both cattle and swine were taken by the fishermen to Newfoundland and Nova Scotia in 1553. In 1604 Lescarbot, a French lawyer, carried cattle to Arcadia. Five years later he wrote a history of New France, wherein he states that in 1508 or thereabouts Baron de Lery attempted a settlement at Sable Island, and the cattle found many years later on the island are supposed to be descendants of De Lery's stock. When the Huguenots settled, in 1672, on the Broad River, S. C., they may have taken cattle with them, but up to the present time we have found no record to that effect.

The first cattle to reach the territory now included in the United States were brought by Sir Richard Grenville to Virginia in 1535, in an expedition sent out by Sir Walter Raleigh, which left Plymouth, England, on April 9, but the colony perished and the cattle were probably slaughtered by the settlers. From an exhaustive study of the records[1] it is evident that the cattle introduced at Jamestown, Va., were from English breeds, with some mixture of Spanish cattle from the West Indies. In New York the cattle were largely of Dutch origin. In Pennsylvania the cattle were brought over by the Dutch and Swedish settlers. At Plymouth, Mass., the cattle were brought from Holland and England. The ships which arrived at Boston contained mostly English breeds, the Devon predominating. In New Hampshire Capt. Mason introduced a large yellow breed from Denmark. In Canada the importations were largely from France.

In the West Indies, Mexico, and Central and South America the cattle were nearly all from Spanish stock until within recent years. Many good breeding animals from improved breeds have been imported from Europe and the United States to Argentina and other countries of South and Central America during the past 50 years. Humped cattle of India (zebus) have also been imported to Texas and the West Indies, because of their supposed immunity from Texas fever, and the crosses with native cows have thus far been very successful.

Although we speak of the different breeds of cattle that were brought to America, there were many nondescripts, and but few if any would be recognized as belonging to any type of our modern purebreds until Messrs. Miller and Gough imported what were supposed to be pure Shorthorns in 1783. With the exception of the Holsteins and the Brown Swiss, but few cattle from other countries have been imported except from the British Islands.

Cattle not being native to America, there are no strictly American breeds, but, owing to differences in climate, care, and ideals of American breeders, the European breeds which have been brought to America have nevertheless changed to some extent. Occasionally a strain of improved stock has arisen as a sport, or by careful selection of the stock has obtained a local reputation as a breed, such as the Gore breed, well known in New England 75 years ago, and still later the American Holderness in New York; but, up to the present time, with the exception of the Polled Durham and the French-Canadian, none of the so-called American breeds has obtained anything like a national reputation.

CLASSIFICATION OF MODERN BREEDS.

Early classifications of breeds were based upon their geographical distribution. Sturm, in 1825, gave the classification of lowland, upland, and mountain breeds. Later, the factor of color was taken

---

[1] The results of a study of the early importations of cattle in America and cattle breeding in colonial times will be given more in detail in a later paper.

into consideration. A half century ago the cattle in Great Britain were divided into shorthorns, middlehorns, and longhorns. To-day, English and American writers arrange them according to their economic value as beef breeds, dairy breeds, and general-purpose cattle, while the Germans make physiographical factors an important consideration, as Krafft, who made seven divisions, namely, steppe, lowland, solid-colored mountain, spotted valley, upland, English, and French breeds. Wagner (1837) was the first to rely upon osteological characters. His two classes were arranged according to the curvature of the vertebral column: (1) *Taurus hypselurus—cauda altissima posita*, and (2) *Taurus frisius—cauda profunda posito*—a questionable division.

Sanson, following the methods of the anthropologists, divided cattle into two groups, the dolichocephalic or long-headed, and the brachycephalic or broad-headed. These two groups he subdivided again, making four subspecies or races. These were: (1) *Bos taurus ligeriensis*, (2) *Bos taurus jurassicus*, (3) *Bos taurus batavicus*, and (4) *Bos taurus alpinus;* the first two being brachycephalic and the two latter dolichocephalic.

1. *Ligeriensis* is found to-day from the mouth of the Loire to that of the Gironde, and in characteristics this race resembles *primigenius*.

2. *Jurassicus* includes *trochoceros* of Meyer and *frontosus* of Nilsson. Its representatives are found in the cattle of Berne and Freiburg, and the breeds in France known as Bressane, Comtoire, Femeline, and Charolais.

3. *Batavicus* is equivalent to *longifrons* of northern Europe.

4. *Alpinus* is the *longifrons* of Switzerland, now generally known as the Brown Swiss breed. What Wilckens calls *brachycephalus*, Sanson would probably consider a cross between *jurassicus* and *batavicus*.

Rütimeyer (1862) was the first to propose a classification based essentially upon zoological characteristics alone, but would include geographical, historical, and geological evidence. It may be called a paleontological classification. His system has been followed more or less closely by all later German authors. He made three races.

1. The primigenius race, which included Holland, Friesland, Oldenburg, Roman, and Podolian breeds, and the white Park cattle of Great Britain.

2. The brachyceros race, which included the Brown Swiss cattle of Switzerland, Uri Wallis, Oberbasle, and Graubünden, and the cattle of Algeria.

3. The frontosus race, which included the spotted cattle of Simmenthal and Saanenthal and the hornless cattle in the mountains of Norway.

The Freiburg breed he considered a cross between the last two races.

Werner (1902) and Wilckens make four groups, *primigenius*, *frontosus*, *brachyceros* (*longifrons*), and *brachycephalus*. Müller (1900) adds a fifth to the four above mentioned and calls it the Highland breed.

Keller's classification (1902 and 1905) is as follows:

1. *Bos primigenius*, whose home is in Europe, includes the English Park cattle, the North German, Lowland, Dutch, Steppe, Simmenthaler, and Freiburg spotted breeds.

2. *Bos sondaicus*, whose original home was in Java, includes the Asiatic and African zebu, Old Egyptian Longhorn, Algerian, marsh cow of the lake dwellers, Albanian, Sardinian, Spanish, Polish, Channel Islands, Hornless Fjell, and Brown Swiss breeds.

Adametz (1898) recognizes two main types of domesticated cattle. From the first, which he calls *Bos taurus primigenius*, and whose ancestor was *Bos primigenius*, he finds four races: (1) Steppe cattle; (2) primigenius mountain cattle, as the breed of Auvergne; (3) primigenius lowland, pure as in Normandy, less pure in the Holland, Oldenburg, and East Friesland red breeds; (4) Swiss spotted or Alpine breeds with a broad forehead (*frontosus*).

His second type is *Bos taurus europeus*, with *Bos brachyceros* (*longifrons*) as the ancestor, represented in five modern races. The first race is typical of the ancestor, and is represented by old marsh cow, the Jersey, Brittany, Illyrian, Albanian, and other similar breeds. The second race is polled; examples are found in Scotland, Sweden, Lapland, and Russia. The third race is of longifrons type, though having become somewhat larger through care and a favorable climate. Brown Swiss is a typical example. The fourth race is represented by Tuxer and Zillerthaler, which are pug-nosed variations of *longifrons*. The fifth race, which he calls pseudo-primigenius, is a result of crossing other types, and is represented by Ayrshire, West Highland, and English Park cattle.

The most ambitious attempt to classify and describe all varieties of cattle is that in Werner's "Rinderzucht," to which the reader is referred for details, it being too long to be given here.

Lydekker (1904) classifies British cattle according to color, making three groups, as follows:

1. Pembroke and Park cattle, either black or white; the black being descendants of forest-dwelling animals, and the white, sports.

2. Spanish and Channel Island cattle, that vary from black to fawn, both primitive colors.

3. Shorthorn, Devon, Hereford, in which black is lost, and being the most marked departure from the primitive types.

A CLASSIFICATION OF BRITISH CATTLE.

McConnell makes three types of cattle in the British Isles, namely, *longifrons*, *primigenius*, and mixed:

Bos longifrons type.

Sutherland.	Anglesea.	Cornish Black.
North Highland.	Carnarvon.	Jersey.
Kintail.	Cardigan	Guernsey.
Kyloe.	Carmarthen.	Alderney.
Skye.	Pembroke.	Irish Moyle.
Galloway.	Merinoneth.	Kerry.
Cumberland.	Brecon Black.	

Bos primigenius type.

Cadzow.	Lincoln Red.	Sussex.
Chillingham.	Lincoln Dutch.	Dorset.
Chartley.	Craven Longhorn.	Glamorgan.
Fife (Falkland).	Derby Longhorn.	Castlemartin Black.
Shorthorn.	Stafford Longhorn.	Castlemartin White.
Teeswater.	Suffolk Dun.	Irish Longhorn.
Holderness.	Hereford.	

Mixed type.

Shetland.	Ayrshire.	Montgomeryshire-
Orkney.	Lothian.	Smokyface.
Banff.	Ettrick.	Old Gloucester.
Aberdeen Horned.	Yorkshire Polled.	North Devon.
Buchan Humlie.	Yorkshire Middlehorn.	South Devon.
Angus Doddie.	Norfolk Horned.	Irish Middlehorn.
Forfar Horned.	Red Polled.	
Argyle (South Highland).	Shropshire.	

In our opinion the Galloway, Glamorgan, Guernsey, and possibly others should be classed with the mixed type.

A CLASSIFICATION OF FRENCH BREEDS.

The following is a classification of French breeds modified from Werner:

Bos primigenius germanicus var. *flandricus*.
    Flamande.
    Ardennaise or Meusienne.
    Wallone.

Bos primigenius germanicus var. *normannus*.
    Normande.

Bos longifrons alpestris var. *flavus*.
    Tarentaise.

Bos longifrons ligeriensis (Sanson).
    Parthenaise (including Poitevine, Mantaise, Vendée, and Marchoise).

Bos longifrons vasconiensis.
    Gasconne.
    Bazadaise.
    D'Aure (St. Giron and Ariegeoise).

Bos frontosus fronconicus.
    Fémeline.
    Charolaise (Nivernaise).
    De Sans (Du Mézenc).
    De Lourdes.

Bos brachycephalus isolanus.
    Carmargue.
    Béarnaise (Basquaise, D'Urt).
    Landaise.

Bos brachycephalus aquitanicus.
    Garonnaise.
    Agenaise.
    Limousine.

Bos brachycephalus celticus.
    Bretonne.

Bos brachycephalus alverniensis.
    De Salers (Du Cantal).
    Du Puy de Dôrne.

CLASSIFICATION OF BREEDS IN AMERICA.

As in England, the breeds of cattle in America are usually classified according to their uses into beef, dairy, and dual-purpose breeds. Hitherto, no attempt has been made to classify them according to their ancestry, but by following Werner's arrangement of types we have the following provisional classification, which includes the breeds which have originated in America as well as those which have been imported. In but few cases, however, are they true to the types, because of much intercrossing.

1. *Bos primigenius germanicus.*
    Holstein-Friesian.
    Dutch Belted.
    Yellow Danish.

  Bos primigenius germanicus var. *normannus.*
    Normandy.

  Bos primigenius germanicus var. *anglo-saxonicus.*
    Shorthorn.
    Polled Durham.
    American Holderness.
    Red Polled.

2. *Bos primigenius scoticus.*
    West Highland.
    Galloway.
    Aberdeen-Angus.
    Ayrshire (crossed with *longifrons*).

3. *Bos longifrons alpestris* var. *brunneus.*
    Brown Swiss.

4. *Bos longifrons isolanus.*
    Jersey.
    Polled Jersey.
    Guernsey (crossed with *primigenius*).

5. *Bos frontosus burgundicus.*
    Simmenthal.
    Mixed breeds from Sweden.
6. *Bos brachycephalus ibericus.*
    Texas.
    Guinea.
7. *Bos brachycephalus celticus.*
    North Wales.
    Kerry.
    Jamestown.
    Brittany.
    French-Canadian.
8. *Bos brachycephalus licestriensis.*
    Longhorn.
9. *Bos brachycephalus britannicus.*
    Hereford.
    Polled Hereford.
    Kansan.
    Sussex.
    Devon.
    Polled Devon.
10. *Bos indicus.*
    Zebu.
11. *Bos bison* ×, *Bos primigenius scoticus.*
    Cattalo.

This arrangement is faulty in many respects and can not be accepted as it now stands. The longhorn breed is presumably of the primigenius type, with possibly some mixture of longifrons blood. The Hereford, Polled Hereford, and Kansan have about the same origin as the Normandy breed. Probably the Ayrshire, West Highland, and Galloway have more longifrons blood than primigenius, while the Kerry and Brittany are more like *longifrons* than *brachycephalus*.

BIBLIOGRAPHY.

Many thousands of articles have been written on the ancestry of domesticated cattle, but it is possible to insert here only a few of the more important articles and such as have been referred to in the preceding pages. A more extensive bibliography is in preparation.

ADAMETZ, L. Untersuchungen über das Rind der Wahima- (Watussi-) Stämme. (Bos Zebu africanus Watussi.) Journal für Landwirtschaft, Jahrgang 42, pp. 137–155. Berlin, 1894.

ADAMETZ, L. Studien über Bos (brachyceros) europaeus, die wilde Stammform der Brachyceros-Rassen des europäischen Hausrindes. Journal für Landwirtschaft, Jahrgang 46, pp. 269–320. Berlin, 1898.

ALLEN, GRANT. Anglo-Saxon Britain. London, 1901.

ARENANDER, E. D. Studien über das ungehörnte Rindvieh im nördlichen Europa unter besonderer Berücksichtigung der nordschwedischen Fjellrasse, nebst Untersuchungen über die Ursachen der Hornlosigkeit. Berichte aus dem physiologischen Laboratorium und der Versuchsanstalt des landwirthschaftlichen Instituts der Universität Halle, Band 3, Heft 13, p. 178. Dresden, 1898.

AULD, R. C. Hornless ruminants. American Naturalist, vol. 21, pp. 730–746; 885–902; 1076–1098. Philadelphia, 1887.

AULD, R. C. The segregations of polled races in America. American Naturalist, vol. 23, pp. 677–686. Philadelphia, 1889.

BAKKER, DIRK L. Studien über die Geschichte, den heutigen Zustand und die Zukunft des Rindes und seiner Zucht in den Niederlanden, mit besonderer kritischer Berücksichtigung der Arbeitsweise des Niederländischen Rindviehstammbuches. 138 pp. Inaugural-Dissertation, Universität Bern, 1909.

BARAŃSKI, ANTON. Geschichte der Thierzucht und Thiermedicin im Alterthum. Wien, 1886.

BARAŃSKI, ANTON. Die vorgeschichtliche Zeit im Lichte der Hausthiercultur. 296 pp. Wien, 1896.

BELTZ, R. Bos primigenius im Mittelalter. Globus, Band 73, no. 7, pp. 116–117. Braunschweig, 1898.

BLYTH, EDWARD. On the animal inhabitants of ancient Ireland. Proceedings of the Royal Irish Academy, vol. 8, pp. 472–476. Dublin, 1864.

BOJANUS, LUDWIG H. De Uro Nostrate Eiüsque Sceleto Commentatio. Verhandlungen der Kaiserlichen Leopoldinisch-Carolinischen Akademie der Naturforscher, Band 13, Abtheilung 2, pp. 413–478. Bonn, 1827.

BOULE, M. Les Grottes de Grimaldi. Geologie et Palentologie. Monaco, 1910, Vol. 1, pt. 3, pp. 157–236.

BREHM, A. E. Die Säugethiere. Band 1–3. Leipzig, 1876–1877.

CATTLE AND DAIRY FARMING. United States Consular Reports. 855 pp. 368 plates. Washington, 1887.

COPE, E. D. The Artiodactyla. American Naturalist, vol. 22, pp. 1079–1095; vol. 23, pp. 111–136. Philadelphia, 1888–1889.

COPE, E. D. The classification and phylogeny of the Artiodactyla. Proceedings of the American Philosophical Society, vol. 24, pp. 377–400. Philadelphia, 1887.

CORNEVIN, Ch. Note sur les bœufs découverts dans les fouilles exécutées Rue de Trion à Lyon-Fourvière, 1885. Bulletin de la Société d'Anthropologie de Lyon, tome 4, pp. 182–187. Lyon, 1885.

DAWKINS, WILLIAM BOYD. On the fossil British oxen. Part 1. Bos Urus, Cæsar. Quarterly Journal of the Geological Society of London. Proceedings. vol. 22, pp. 391–402. London, 1866.

DAWKINS, WILLIAM BOYD, and SANFORD W. AISHFORD. British Pleistocene Mammalia. Parts 1–6. London, 1866–1887.

DÜRST, J. ULRICH. Animal remains from the excavations at Anau, and the horse of Anau in its relation to the races of domestic horses. Explorations in Turkestan. Expedition of 1904. vol. 2, pp. 341–344. Washington, 1908.

DÜRST, J. ULRICH. Die Rinder von Babylon, Assyrien und Ägypten und ihr Zusammenhang mit den Rindern der alten Welt . . . 94 pp. Berlin, 1899.

EWART, J. C. Hornless cattle. Live Stock Journal, vol. 70, pp. 599–600. London, Dec. 3, 1909.

EWART, J. C. On skulls of oxen from the Roman military station at Newstead, Melrose. Proceedings of the Zoological Society of London, 1911, part 2, pp. 249–282.

FALCONER, HUGH. Descriptive Catalogue of the Fossil Remains of Vertebrata from the Sewalik Hills, the Nerbudda, Perim Island, etc., in the Museum of the Asiatic Society of Bengal. 261 pp. Calcutta, 1859.

FALCONER, HUGH. Palæontological Memoirs and Notes. 2 vols. London, 1868.

FRAAS, E. Römische Statuetten von Wisent und Ur. Fundberichte aus Schwaben (Württembergischer Anthropologischer Verein), Band 7, p. 37. Stuttgart, 1899.

Frantzius, A. v. Die Urheimath des europäischen Hausrindes. Archiv für Anthropologie, Band 10, pp. 129–137. Braunschweig, 1878.

Gadow, Hans. The Evolution of Horns and Antlers. Proceedings of the . . . Zoological Society of London, 1902, vol. 1 (Jan.–April), pp. 206–222.

Gesner, Conrad. Historiae Animalium Lib. I. de Quadru pedibusviviparis. Tiguri, 1551.

Gray, J. E. Bos brachyceros, the West-African Buffalo, and the Dwarf Buffalo of Pennant. Annals and Magazine of Natural History . . . vol. 12 (4th series), pp. 499–500. London, 1873.

Gray, J. E. Descriptions of some new or little known mammalia, principally in the British Museum Collections. Magazine of National History, vol. 1, new series, pp. 577–587. London, 1837.

Gregory, William K. The Orders of Mammals. Bulletin of the American Museum of Natural History, vol. 27. 524 pp. New York, 1910.

Griffith, Edward. Animal Kingdom . . . by the Baron Cuvier, with additional descriptions . . . vol. 4, p. 416. London, 1827.

Hahn, Eduard. Die Haustiere und ihre Beziehungen zur Wirtschaft des Menschen. Eine geographische Studie. 581 pp. Leipzig, 1896.

Hartmann, Robert. Die Haussäugethiere der Nilländer. Annalen der Landwirthschaft in den Königlich Preussischen Staaten, Band 43, pp. 281–310; Band 44, pp. 7–38. Berlin, 1864.

Hartmann, Robert. Studien zur Geschichte der Hausthiere. Zeitschrift für Ethnologie . . . Band 1, pp. 66, 232, 353. Berlin, 1869.

Hay, Oliver Perry. Bibliography and catalogue of the fossil vertebrata of North America. United States Geological Survey, Bulletin 179. Washington, 1902.

Hengeveld, G. J. Het Rundvee, zijne verschillende soorten, rassen en veredeling. Ed. 2, vol. 1, 467 pp.; vol. 2, 519 pp. Haarlem, 1865.

Hengéveld, G. J. [On the origin of the "Holstein Breed."] Report of the Commissioner of Agriculture, 1874, pp. 403–407. Washington, 1875.

Herberstein, Sigmund, Freiherr von. Rerum Moscoviticarum Commentarii . . . Basileae, 1549. (Later editions were given different titles.)

Hilzheimer, Max. Wisent und Ur im K. Naturalienkabinett zu Stuttgart. Jahreshefte des Vereins für vaterländische Naturkunde in Württemberg, Jahrgang 65, pp. 241–269. Stuttgart, 1909.

Huet, J. Les Bovidés. Bulletin Bimensuel de la Société Nationale d'Acclimatation de France (Revue des Sciences Naturelles Appliquées), 1er semestre, année 38, pp. 1; 334. Paris, 1891.

Hughes, T. McKenny. The Evolution of the British Breeds of Cattle. Journal of the Royal Agricultural Society of England, vol. 5 (series 3), pp. 561–563. London, 1894.

Hughes, T. McKenny. On the more important breeds of cattle which have been recognised in the British Isles in successive periods, and their relation to other archaeological and historical discoveries. Archaeologia . . . vol. 5 (2d series), pp. 125–158. London, 1896.

Jarocki, E. P. v. Zubr oder der lithauische Auerochs. Hamburg, 1830.

Kaltenegger, Ferd. Die geschichtliche Entwickelung der Rinderrassen. Correspondenz-Blatt der Deutschen Gesellschaft für Anthropologie, Ethnologie und Urgeschichte. Jahrgang 25, no. 9, p. 121; no. 10, p. 123. München, 1894. Also in: Mittheilungen der Anthropologischen Gesellschaft in Wien, Sitzungsberichte, Band 24, Number 4, pp. 111–116. Wien, 1894.

Kaltenegger, Ferd. Iberisches Hornvieh in den Tiroler und Schweizer Alpen. Mittheilungen der Anthropologischen Gesellschaft in Wien, Band 14 (neue Folge, Band 4), pp. 129–141. Wien, 1884.

KALTENEGGER, FERD. Die geschichtliche Entwicklung der Rinderracen in den österreichischen Alpenländern. Jahrbuch für österreichische Landwirthe. Prag, 1881.

KALTENEGGER, FRED. Die österreichischen Rinderracen. Parts 1–6. Vienna, 1879–1899.

KELLER, CONRAD. Die Afrikanischen Elemente in der europäischen Haustierwelt. Globus, Band 72, pp. 285–289. Braunschweig, November 6, 1897.

KELLER, CONRAD. Die Haustierwelt Asiens. Westermanns Illustrierte Deutsche Monatshefte . . . Jahrgang 43, Band 86, pp. 567–575. Braunschweig, 1899.

KELLER, CONRAD. Die Abstammung der ältesten Haustiere. 232 pp. Zürich, 1902.

KELLER, CONRAD. Naturgeschichte der Haustiere. 304 pp. Berlin, 1905.

KELLER, CONRAD. Nochmalsie Goldbecher von Vaphio. Globus, vol. 74, pp. 81–82. Braunschweig, 1898.

KITT, TH. Ueber Kulturformen von Bos brachyceros. (Als Beitrag zur natürlichen Geschichte der bayrischer Rinderrassen.) Landwirthschaftliche Jahrbücher, Band 13, pp. 593–614. Berlin, 1884.

KÖHLER, E. M. Haustiere der Chinesen. Der Zoologische Garten . . . Jahrgang 41, pp. 5; 33; 65. Frankfurt a. M., 1900.

KRÄMER, HERMANN. Die Haustierfunde von Vindonissa . . . Revue Suisse de Zoologie, tome 7, pp. 143–272. Genève, 1899.

KRAFFT, GUIDO. Thierzuchtlehre. Berlin, 1900.

KRAUSE, ERNST H. L. Zur Würdigung der alten Abbildungen europäischer Wildrinder. Globus, Band 73, no. 24, pp. 389–390. Braunschweig, 1898.

LANGKAVEL, G. Die Verbreitung des Hausrindes in Nordafrika auser Algerien. ([auch] in Südafrika.) Zeitschrift für Wissenschaftliche Geographie, Band 4, pp. 16–28; Band 5, pp. 172–186. Wien, 1883–4.

DE LAPPARENT. Étude sur les races, variétés et croisements de l'espèce bovine en France. Annales Ministère de l'Agriculture [France], année 21, no. 1, pp. 177–243; no. 2, pp. 322–394. Paris, 1902.

LORTET and GAILLARD, C. La Faune momifiée de l'Ancienne Égypte. Archives du Muséum d'Histoire Naturelle de Lyon, tome 8, mémoire no. 2, 206 pp. Lyon, 1903.

LYDEKKER, RICHARD. Crania of Ruminants from the Indian Tertiaries. Memoirs of the Geological Survey of India. Palaeontologia Indica, series 10, vol. 1, pp. 88–181. Calcutta, 1880.

LYDEKKER, RICHARD. [Fossil Vertebrata of India.] Records of the Geological Survey of India, vol 10, p. 30; vol. 16, p. 61; vol. 20, p. 51. Calcutta, 1877, 1883, 1887.

LYDEKKER, RICHARD. Wild Oxen, Sheep, and Goats of all Lands, Living and Extinct. 318 pp. London, 1898.

LYDEKKER, RICHARD. Colour of Cattle. The Field, the Farm, the Garden, vol. 104, p. 181. London, July 23, 1904.

McCONNELL, PRIMROSE. Elements of Agricultural Geology . . . 329 pp. London, 1902.

MAJOR, C. J. FORSYTH. On the Mammalian Fauna of the Val d'Arno. Quarterly Journal of the Geological Society of London, vol. 41, pp. 1–8. London, 1885.

MAJOR, C. J. FORSYTH. Note on a Pliocene Mammalian Fauna at Olivola in the Upper Val di Magra (Prov. Massa-Carrara), Italy. Geological Magazine . . . vol. 7 (new series, decade 3), pp. 305–308. London, 1890.

MIDDENDORFF, A. VON. Ueber die Rindviehrasse des nördlichen Russlands und ihre Veredlung. Landwirthschaftliche Jahrbücher, Band 17, pp. 267–327. Berlin, 1888.

Morse, Elisha W. Some early history concerning the Shorthorn Breed in New England. New England Farmer, vol. 87, Feb. 29, p. 8. Brattleboro, 1908.

Mortillet, Gabriel de. Origines de la Chasse, de la Pêche et de l'Agriculture. Première partie. Paris, 1890.

Müller, Robert. Grundzüge der landwirtschaftlichen Tierproduktionslehre. 439 pp. Berlin, 1900.

Müller, Sophus. Nordische Altertumskunde nach Funden und Denkmälern aus Dänemark und Schleswig. Translated from the Danish by Otto L. Jiriczek. 2 vols. Strassburg, 1897–1898.

Nathusius, Hermann v. Ueber Schädelform des Rindes. Landwirthschaftliche Jahrbücher, Band 4, pp. 441–459. Berlin, 1875.

Nehring, A. Die Herberstain'schen Abbildungen des Ur und des Bison. (Ein Beitrag zur Geschichte des europäischen Urrindes.) Landwirthschaftliche Jahrbücher, Band 25, pp. 915–933. Berlin, 1896.

Nehring, A. Einige Bemerkungen über Anton Wieds "Moscovia" und das zugehörige Urusbild. Globus, Band 71, pp. 85–89. Braunschweig, January 30, 1897.

Nehring, A. Einige Bemerkungen über Anton Wieds "Moscovia" und das zugehörige Urusbild. Globus, Band 71, pp. 85–89. Braunschweig, Jan. 30, 1897.

Nehring, A. Ein Gedicht über Ur und Bison aus dem Jahre 1552. Globus, Band 71, pp. 242–243. Braunschweig, April 10, 1897.

Nehring, A. Über den Einfluss der Domestication auf die Grösse der Thiere, namentlich über Grössenunterschiede zwischen wilden und zahmen Grunzochsen (Poëphagus grunniens). Sitzungs-Bericht der Gesellschaft naturforschender Freunde zu Berlin, pp. 133–145. Berlin, 1888.

Nehring, A. Über Herberstain und Hirsfogel ... Berlin, 1897.

Nilsson. On the extinct and existing Bovine Animals of Scandinavia. Annals and Magazine of Natural History ... vol. 4 (series 2). pp. 256–269. London, 1849.

Osborn, Henry F. The Age of Mammals ... 635 pp. New York, 1910.

Owen, Richard. History of British Fossil Mammals and Birds. 560 pp. London, 1846.

Owen, Richard. Paleontology, or a systematic summary of extinct animals and their geological relations. Edinburgh, 1860.

Owen, Richard. Report on the British Fossil Mammalia. Report of the 13th meeting of the British Association for the Advancement of Science, pp. 208–241. London, 1844.

Pagenstecher. Studien zum Ursprung des Rindes, mit einer Beschreibung der fossilen Rinderreste des Heidelberger Museums. Fühling's landwirthschaftliche Zeitung, Jahrgang 27. Berlin und Leipzig, 1878.

Pallas, P. S. De ossibus Sibiriae fossilibus craniis praesertim Rhinocerotum atque Buffalorum, observationes. Novi Commentarii Academiae Scientiarum Imperialis Petropolitanae, tome 13, pp. 436–477. Petropoli, 1769.

Palmer, T. S. Index Generum Mammalium: a list of the genera and families of mammals. U. S. Department of Agriculture, Biological Survey. North American Fauna, no. 23. Washington, 1904.

Pomel, Auguste. Catalogue méthodique et descriptif des Vertébrés fossiles ... Paris, 1853.

Pusch, E. G. Die Beurteilungslehre des Rindes. 388 p. Berlin, 1896.

Pusch, G. Neue Beiträge zur Erläuterung und endlichen Erledigung der Streitfrage über Tur und Zubr. (Urus und Bison.) Archiv für Naturgeschichte, Jahrgang 6, Band 1, pp. 47–137. Berlin, 1840.

RASMUSSEN, FREDERIK. Cattle breeders' associations in Denmark. U. S. Department of Agriculture, Bureau of Animal Industry. Bulletin 129. 40 pp. Washington, 1911.

REGEL, A. Die einheimischen und angebauten Kulturpflanzen des oberen Amudaria. Gartenflora, Band 33, p. 46. Stuttgart, 1884.

RHOADS, SAMUEL N. Notes on Living and Extinct Species of North American Bovidae. Proceedings of the Academy of Natural Sciences of Philadelphia, 1897, pp. 483–502. Philadelphia, 1898.

RINGHOLZ, P. O. The history of cattle breeding at Einsiedeln. Landwirtschaftliche Jahrbuch de Schweiz, Jargang 22, Heft 8, pp. 413–508. Bern, 1908.

RÜTIMEYER, L. Die Fauna der Pfahlbauten in der Schweiz. Neue Denkschriften der allgemeinen Schweizerischen Gesellschaft für die gesammten Naturwissenschaften. Band 19, pp. 1–248. Zürich, 1862.

RÜTIMEYER, L. Die Rinder der Tertiär-Epoch nebst Vorstudium zu einer natürlichen Geschichte der Antilopen. Abhandlungen der Schweizerische Paläontologische Gesellschaft. Zürich, 1877–1878.

RÜTIMEYER, L. Versuch einer natürlichen Geschichte des Rindes in seinen Beziehungen zu den Wiederkäuern im Allgemeinen. pp. 102–175. Zürich, 1867.

SANSON, A. Détermination spécifique des ossements fossiles ou anciens de bovidés. Comptes Rendus des Séances de l'Académie des Sciences, tome 87, pp. 756–759. Paris, July–December, 1878.

STUDER, THEOPHIL. Entwickelung der Hausthierzucht bei den Pfahlbauern. Correspondenz-Blatt der deutschen Gesellschaft für Anthropologie, Ethnologie und Urgeschichte, Jahrgang 30, pp. 172–174. München, November and December, 1899. Also in: Mittheilungen der Anthropologischen Gesellschaft in Wien, Band 30 (neue Folge, Band 20), pp. 106–108. Wien, January, 1900.

STUDER, THEOPHIL. Die Thierwelt in den Pfahlbauten des Bielersees. Mittheilungen der Naturforschenden Gesellschaft in Bern, 1882. Heft 2, Nummer 1042, pp. 17–115. Bern, 1883.

STUDER, THEOPHIL. Ueber Säugethierreste aus glacialen Ablagerungen des bernischen Mittellandes. Mittheilungen der Naturforschenden Gesellschaft in Bern, 1888, pp. 66–70. Bern, 1889.

SUNDBÄRG, GUSTAV. Sweden; its people and its industry. 1141 pp. Stockholm, 1904.

SWINHOE, ROBERT. Catalogue of the Mammals of China (south of the River Yangtsze) and of the Island of Formosa. Proceedings of the ... Zoological Society of London, 1870, pp. 615–653.

THOMAS, PHILIPPE. Recherches stratigraphiques et paléontologiques sur quelques formations d'eau douce de l'Algérie. Mémoires de la Société Géologique de France, tome 3 (série 3), 2, pp. 1–51. Paris, 1884.

THOMAS, PH. Recherches sur les bovidés fossiles de l'Algérie. Bulletin de la Société Zoologique de France, vol. 6, pp. 92–136, plate 3. Paris, 1881.

TRÖLTSCH, E. VON. Die Pfahlbauten des Bodenseegebietes. 255 pp. Stuttgart, 1902.

TROUESSART, EDOUARD L. Catalogus Mammalium tam Viventium quam Fossilium. Nova Editio (Prima completa). Vols. 1–2. 1469 pp. Berlin, 1897–1899.

——— ——— Quinquennale Supplementum. 929 pp. Berlin, 1904–1905.

WAGNER, ANDREAS. Naturgeschichte des Rindes. Erlangen, 1837. (Abdruck aus: Schreber's Säugethiere, Band 5.)

WERNER, HUGO. Ein Beitrag zur Geschichte des europäischen Hausrindes. 48 pp. Berlin, 1892.

WERNER, HUGO. Die Geschichte des europäischen Hausrindes in ihren Beziehungen zu den Völkern Europas. Deutsche Landwirtschaftliche Presse, Jahrgang 23, pp. 516–517. Berlin, 1896.

WERNER, HUGO. Die Rinderzucht . . . 645 pp. 136 plates. Berlin, 1892. Second edition, 1902.

WILCKENS, MART. Die abändernden Einflüsse der Kultur auf die Form des Rinderschädels. Landwirthschaftliche Jahrbücher, Band 5, pp. 651–653. Berlin, 1876.

WILCKENS, MART. Ueber die Schädelknochen des Rindes aus dem Pfahlbau des Laibacher Moores. Mittheilungen der Anthropologischen Gesellschaft in Wien, Band 7, 1877, pp. 165–175. Wien, 1878. Band 9, 1880.

WILCKENS, MART. Grundzüge einer Naturgeschichte des Hausrindes. Vienna, 1885.

WILSON, JAMES. The Evolution of British cattle and the fashioning of breeds. 147 pp. London, 1909.

WILSON, JAMES. The Scandinavian origin of the hornless cattle of the British Isles. Scientific Proceedings of the Royal Dublin Society, vol. 12 (new series), no. 15, pp. 145–164. Dublin, 1909.

WOLLEMANN, A. Ein domestizirtes Zwergrind der Primigeniusrasse. Correspondenz-Blatt der deutschen Gesellschaft für Anthropologie, Ethnologie und Urgeschichte, Jahrgang 22, pp. 50–51. München, July, 1891.

WOODWARD, ARTHUR S. Outlines of Vertebrate Palaeontology for Students of Zoology. 470 pp. Cambridge, England, 1898.

WRZESNIOWSKI, AUGUST. Studien zur Geschichte des polnischen Tur (Ur, Urus, Bos primigenius Bojanus). Zeitschrift für wissenschaftliche Zoologie, Band 30, pp. 493–555. Leipzig, 1878.

YULE, HENRY. Travels of Marco Polo . . . Edition 2. 2 vols. London, 1875.

ZITTEL, KARL A. Handbuch der Palaeontologie. 1. Abtheilung, Band 4, Vertebrata (Mammalia). München und Leipzig, 1891–1893.

www.ingramcontent.com/pod-product-compliance
Lightning Source LLC
Chambersburg PA
CBHW062229220526
45471CB00009B/3406